Lecture Notes in Mathematics   1612

Editors:
A. Dold, Heidelberg
F. Takens, Groningen

T0255419

Springer
Berlin
Heidelberg
New York
Barcelona
Budapest
Hong Kong
London
Milan
Paris
Tokyo

Wolfgang Kühnel

Tight Polyhedral Submanifolds and Tight Triangulations

 Springer

Author

Wolfgang Kühnel
Mathematisches Institut B
Universität Stuttgart
Pfaffenwaldring 57
D-70550 Stuttgart, Germany
E-mail: kuehnel@mathematik.uni-stuttgart.de

Cataloging-in-Publication Data applied for

Die Deutsche Bibliothek - CIP-Einheitsaufnahme

Kühnel, Wolfgang:
Tight polyhedral submanifolds and tight triangulations /
Wolfgang Kühnel. - Berlin ; Heidelberg ; New York ;
Barcelona ; Budapest ; Hong Kong ; London ; Milan ; Paris ;
Tokyo : Springer, 1995
  (Lecture notes in mathematics ; 1612)
  ISBN 3-540-60121-X
NE: GT

Mathematics Subject Classification (1991): 53C42, 52B70, 57Q15, 57Q35

ISBN 3-540-60121-X Springer-Verlag Berlin Heidelberg New York

© Springer-Verlag Berlin Heidelberg 1995
Printed in Germany

Typesetting: Camera-ready TeX output by the author
SPIN: 10479510   46/3142-543210 - Printed on acid-free paper

Preface

A first version of this monograph was written several years ago, while the author was a guest of the I.H.E.S. at Bures-sur-Yvette. He gratefully acknowledges numerous discussions about tight submanifolds with N.H. Kuiper, T.F. Banchoff and others. Later this first draft was distributed as preprint Nr. 108 of the Math. Dept., University of Duisburg. The author also acknowledges the warm hospitality of the Landau-Center at the Hebrew University of Jerusalem in 1990. At that time Gil Kalai introduced the author to some of the mysteries of the Upper and Lower Bound Conjecture for polytopes, and certain parts of Chapter 4 were being developed. Fortunately or unfortunately, there are still a number of conjectures left open.

Tightness is a concept from differential geometry that has many connections to other branches of mathematics. This monograph is a presentation of a part of mathematics sitting between various special disciplines such as differential geometry, topology, theory of convex polytopes, combinatorics. The main intention is to stimulate further fruitful interaction in this direction. The treatment of the 2-dimensional case in the – essentially self-contained – Chapter 2 is an example of an interplay between the theory of convex polytopes, graph theory, and elementary polyhedral topology.

Finally, the author should like to thank S. Lukas for typing the first preprint version, B. Dunkel for typing the main part of the present version in LaTeX (including numerous complicated formulas), Ch. Habbe for valuable support in drawing the figures, and D. Cervone and M. van Gemmeren for careful proof-reading and support in handling the LaTeX system.

Only a few days before the manuscript was finished, we received the sad news that Professor Nicolaas Hendrik Kuiper passed away at the age of 74. He was not only a great mathematician and one of the pioneers in this area of mathematics, but also a longstanding friend. This volume is dedicated to his memory.

January 1995

W.K.

Contents

1. Introduction and basic notions

Tightness is a generalization of the notion of convexity that applies to objects other than topological balls and their boundary surfaces. In some sense it means that a submanifold or subset of E^d is embedded 'as convexly as possible' with respect to its topological properties.

The subject of tight embeddings began in 1938 with a paper of A.D. Alexandrov [A] on 'T-surfaces', studying the total absolute curvature for analytic torus-shaped surfaces in three-space. A T-surface in Alexandrov's sense is a torus with minimal total absolute curvature 8π. In 1950, J. Milnor [Mir] proved an inequality between the total absolute curvature of a hypersurface in Euclidean space and its total Betti number. This paper was not published until recently but it anticipated several developments of the subsequent years. Not much later, in 1957, S.S. Chern and R.K. Lashof [CL] initiated the study of total absolute curvature for smooth submanifolds of higher dimension and codimension in Euclidean space, and they proved the inequality between the total absolute curvature and the total Betti number in general. This was done by relating this question to numbers of critical points of height functions restricted to the submanifold. This aspect of tightness was further developed by N.H. Kuiper in a series of papers beginning in 1958 [Kui1].

N.H. Kuiper was the first one to interpret minimal total absolute curvature for non-smooth submanifolds, by extending the critical point theory in an appropriate way. In 1965, T.F. Banchoff [Ba1] introduced tightness and the Two-piece-property for polyhedral surfaces in higher dimensional space. This has been extended by the author to a variety of other contexts, including higher dimensional polyhedra and the relationship of tightness with the classical theory of convex polytopes.

After the early Lecture Notes by D. Ferus [Fe], the monograph by T. Cecil and P. Ryan [CR] gives the background of the theory of tightness for smooth submanifolds including related subjects such as tautness and isoparametric hypersurfaces. In this monograph, we will present the theory of tightness for polyhedral manifolds and their embeddings into Euclidean space, and the related subject of tight triangulations. These are triangulations such that any simplexwise linear mapping into any Euclidean space is tight.

In Chapter 2 we start with the following quite elementary problem: Is it possible to embed compact and connected surfaces (without boundary) as polyhedra into Euclidean space such that any hyperplane separates the surface into at most two pieces? This is called the two-piece-property; in this special case it is equivalent to tightness. We have to distinguish between the case of small codimension 1,2,3 (Section 2B) and higher codimension (Section 2C); in fact, the methods are quite different. It was the idea of T.F. Banchoff to construct tight polyhedral surfaces as subcomplexes of higher dimensional cubes or simplices. In this case the crucial condition is that the surface contains the complete 1-dimensional skeleton of the cube or simplex (or any polytope in general). We shall improve Banchoff's results by applying several tools from the theory of convex polytopes. With a few exceptions, it is always possible to find a tight polyhedral surface $M \to E^d$ for arbitrary given M and d subject to just one essential condition: the Heawood inequality between d and $\chi(M)$. The origin of this inequality is the Map Color Problem for surfaces of genus greater than zero, studied by Heawood at the end of the 19th century. Tight triangulations of surfaces are characterized by the case of equality in the Heawood inequality (Section 2D). Finally, the cases of surfaces with boundary and with singularities are discussed in Section 2E and Section 2F.

Chapter 3 gives the general definition and basic facts about tightness of higher dimensional polyhedra. Here it is quite natural to take advantage of a certain critical point theory which generalizes the classical Morse theory and which is adapted to the polyhedral case. With respect to this critical point theory, we follow the ideas and results of M. Morse and N.H. Kuiper. Section 3C presents a general method how to construct tight subcomplexes of higher dimensional cubes. This is an application of L. Danzer's construction of a 'power-complex' 2^K for a given simplicial complex K. It turns out that 2^K is always tight in the ambient Euclidean space, independently of K. This leads to unexpected types of examples.

Chapter 4 is the central chapter of this monograph. It deals with tightly embedded $(k-1)$-connected $2k$-manifolds, as a natural generalization of the results of Chapter 2 to the case of arbitrary k. Several versions of generalized Heawood inequalities are given. In particular, as in the case of surfaces, the case of equality is exactly the case of tight triangulations: subcomplexes of the simplex containing its complete k-dimensional skeleton. Special examples are the unique 9-vertex triangulation of CP^2 and three different 15-vertex triangulations of an 8-manifold. Various conjectures are made, and various open problems are mentioned.

Chapter 5 deals with the odd-dimensional case. Here it seems that there is no easy way to transform the tightness into a purely combinatorial condition. Therefore the results are much weaker than the ones in Chapter 4. However, there are nontrivial examples. In particular there is a 9-vertex triangulation of a nonorientable handle found by D. Walkup and independently by A. Altshuler and

L. Steinberg. This leads to a tight embedding of the 3-dimensional Klein bottle into E^8. Moreover, for any dimension d there is a tight polyhedral embedding of the sphere product $S^1 \times S^{d-1}$ or the corresponding twisted product into $(2d+2)$-space by a tight triangulation.

Chapter 6 describes a method how to embed connected sums tightly into Euclidean space if the summands admit appropriate tight embeddings. This is a generalization of Banchoff's construction of a tight Klein bottle from two copies of a tight Möbius band. In particular it leads to tight embeddings of connected sums of $\mathbf{C}P^2$ into E^8. This problem is closely related with the discussion of tight embeddings of manifolds with boundary.

Chapter 7 gives some generalizations to the case of odd-dimensional manifolds and pseudomanifolds with isolated singularities. We obtain an inequality of the Heawood type which in a certain sense is an odd-dimensional analogue of the one in Chapter 4. This inequality deals with $(k-1)$-connected $(2k-1)$-pseudomanifolds with isolated singularities, and once again the case of equality coincides with the case of tight triangulations. Slices by hyperplanes are tight polyhedral manifolds if the hyperplane does not meet any of the vertices.

Throughout this volume (except for Chapter 2 which is essentially selfcontained), the following basic notions will be used:

1.1 Definition (Polytopes): A convex d-polytope is the convex hull of finitely many points in E^d not lying in a common hyperplane. This has the structure of a complex built up by the partially ordered set of faces of dimension $0, 1, \ldots, d-1$. The faces are defined as the intersections of the convex set with supporting hyperplanes. The $(d-1)$-dimensional faces are often called facets, the 1-dimensional faces are the edges, the 0-dimensional faces are the vertices. The d-simplex $\triangle^d$ is the convex hull of $d+1$ points in general position in E^d. A d-polytope is called simplicial if for any k each of its k-dimensional faces is a k-simplex; it is called simple if its dual is simplicial, or equivalently, if at each vertex there meet exactly d edges, the minimum number. Each face of a simple polytope is also simple. For each vertex of a convex d-polytope, the vertex figure is defined as the $d-1$-polytope which occurs as a slice by a hyperplane through the d-polytope, separating this vertex from all the other vertices.
The k-dimensional skeleton (or k-skeleton) of a polytope P, denoted by $Sk_k(P)$, is the union of all k-dimensional faces of P. In particular $Sk_1(\triangle^{n-1})$ is combinatorially equivalent to the complete graph with n vertices, denoted by K_n. $Sk_k(\triangle^{n-1})$ is also called the complete k-complex with n vertices, see [Du].

1.2 Definition (Polyhedra): A polyhedron is a finite union of convex polytopes of arbitrary dimensions (called faces) such that the intersection of any two of them is either empty or a common face of both of them. A special case is a

simplicial complex which can always be regarded as a subcomplex of the boundary complex of a simplex of sufficiently high dimension. Sometimes one wants
to allow also non-convex faces. By subdivision one can always reduce this case
to the case of convex faces. However, note that we have to distinguish between
polyhedra and polytopes even in the convex case: the ordinary cube consisting of six squares is a 3-polytope (= the convex hull of its eight vertices). A
subdivided cube consisting of twelve triangles, each square subdivided by a di-
agonal, is a convex polyhedron but not a polytope in this sense. More precisely,
the diagonals are considered as edges of the polyhedron but not as edges of the
polytope. In the literature such distinction is also made between so-called proper
and improper edges or vertices. Every triangulated d-sphere with at most $d + 4$
vertices is the boundary complex of a certain $(d + 1)$-polytope (see [M]), but
there are triangulated 3-spheres with 8 vertices which are non-polytopal (see
[Bar2], [Grü-S]).

1.3 Definition (Tightness): A compact and connected subset $M \subset E^d$ is called
tight with respect to a field F if for every open or closed half space $h \subset E^d$ the
induced homomorphism

$$H_*(h \cap M; F) \to H_*(M; F)$$

is injective where H_* denotes an appropriate homology theory. For polyhedra M
it is convenient to use singular homology or simplicial homology because $h \cap M$
has always the homotopy type of a finite polyhedron. M is called tight if it is
tight with respect to at least one field, where the standard case in the literature
is the field $\mathbf{Z}_2$ with two elements. M is called substantial in E^d if it is not
contained in any hyperplane.

1.4 Definition (top-sets, Tight triangulations): Let M be a polyhedron in
E^d and $\mathcal{H}$ its convex hull, regarded also as a convex polytope. For each k-
dimensional face A_k of $\mathcal{H}$ we call $M \cap A_k$ a k-top-set of M. A top-set is a
k-top-set for some k. Moreover, if M is tight then every top-set of M is also
tight because for any half space h the intersection $h \cap M \cap A_k$ is a deformation
retract of $h' \cap M$ for some nearby half space h'. A top-set is called essential if it
is not a convex set.

If in addition M is a subcomplex of $\mathcal{H}$ then we call it a tight complex. This
carries a structure analogous to that of the d-polytope $\mathcal{H}$ if we replace the k-
faces of $\mathcal{H}$ by the tight k-top-sets of M: it is just the complex built up by the tight
top-sets. A subcomplex M of the boundary complex of a d-dimensional simplex
is called a tight simplicial complex if M is tight regarded as a polyhedron in the
ambient Euclidean space. In this case every top-set is again a tight simplicial
complex.
If it is in addition a triangulation of a manifold with or without boundary, then
we call this simplicial complex a tight triangulation of the manifold.

1.5 Definition (combinatorial manifolds): If the underlying set of a simplicial complex is a topological manifold then it is usually called a triangulated manifold. However, in general these two structures (topological and combinatorial) need not be compatible. There is the strange example of the so-called Edwards sphere [Ed] which is a double suspension of an homology 3-sphere and which is not PL. A PL structure on a manifold is an atlas of charts which are compatible with each other by piecewise linear coordinate transformations. Therefore, a slightly stronger notion is introduced as follows: A simplicial complex is called a combinatorial manifold of dimension d if the link (or vertex figure) of any k-dimensional simplex is a triangulated $(d - k - 1)$-dimensional sphere. This condition implies that this manifold carries a PL structure and that each vertex star is a neighborhood of the corresponding vertex and that this vertex star is PL homeomorphic to a d-ball. Conversely, every PL manifold admits a triangulation which is a combinatorial manifold in this sense. For details see [Hud], [Kui7], [Scht], [RS]. As usual, $\chi(M)$ denotes the Euler characteristic of M.

2. Tight polyhedral surfaces

This chapter is essentially self-contained. It deals with the 2-dimensional case where tightness is a quite elementary property. A <u>surface</u> is a compact and connected 2-dimensional manifold without boundary, unless stated otherwise. At the end of this chapter we consider surfaces with boundary also. A surface M is <u>embedded</u> into E^d if it is homeomorphic to a subset of E^d. Similarly, M is <u>immersed</u> into E^d by a continuous mapping $f\colon M \to E^d$ if f is locally injective. Roughly speaking, 'immersion' means 'local embedding'. A <u>polyhedral surface</u> $M \subset E^d$ is a surface embedded in E^d such that M is a finite union of planar polygons (or <u>faces</u>) where any two of them have no interior parts in common. By subdivision one can always assume that the polygons are convex. Then the intersection of two such convex polygons is either empty or one point (a <u>vertex</u>) or a line segment (an <u>edge</u>), and that the intersection of two edges is either empty or one vertex. Similarly, one can talk about <u>polyhedral immersions</u> of surfaces. Note that in this case the self-intersections of faces are not automatically considered as edges or vertices.

A polyhedral surface can be regarded as the partially ordered set of its vertices, edges and convex faces. We call it <u>triangulated</u> if all faces are triangles. An <u>abstract triangulation</u> of an abstract surface is a decomposition into triangles such that any two of them meet along a vertex or an edge, or not at all. Such an abstract triangulation can always be regarded as a polyhedron in a Euclidean space of higher dimension. $\chi(M)$ denotes the Euler characteristic of M.

2A. 0-tightness

2.1 Definition: A surface $M \subset E^d$ (or an arbitrary connected polyhedron in E^d) is called <u>0-tight</u> if for any open or closed half space $h \subset E^d$ the intersection $M \cap h$ is connected (Banchoff's <u>Two-piece-property</u>, TPP). For an immersion $f\colon M \to E^d$ the 0-tightness means that for any h the preimage $f^{-1}(M \cap h)$ is connected. A subset (or an immersion) is called <u>substantial</u> if it (or its image, respectively) is not contained in any hyperplane. For a compact polyhedron M in E^d let $\mathcal{H} = \mathcal{H}(M)$ denote the <u>convex hull</u> of M, the smallest convex set which contains M. $\mathcal{H}(M)$ is a convex d-polytope if M is substantial in E^d, compare 1.1.

The following necessary condition for the 0-tightness of a polyhedron is originally due to T. Banchoff. In his terminology the 1-skeleton of the convex hull is called the set of extreme vertices and extreme edges.

2.2 Lemma (T. Banchoff [Ba1;Lemma 3.1]): *Let $M \subset E^d$ be a 0-tight and connected polyhedron. Then M contains the 1-skeleton of its convex hull:*

$$Sk_1(\mathcal{H}) \subset M.$$

PROOF: Let e be an edge of $\mathcal{H}$ with endpoints v, w. By construction M contains v and w. There is a half space h of E^d such that $h \cap \mathcal{H} = e$. Consequently we have $\{v, w\} \subset h \cap M \subset h \cap \mathcal{H} = e$. By the 0-tightness $h \cap M$ must be connected. It follows that $h \cap M = e$.

2.3 Lemma: *A polyhedral surface M with convex faces is 0-tight if and only if its 1-skeleton is 0-tight. The same holds for any connected polyhedron M in the sense of Definition 1.2.*

PROOF: If the 1-skeleton is 0-tight then M is 0-tight because adding higher dimensional faces preserves the connectedness of $M \cap h$. Vice versa, if $M \cap h$ is connected then $Sk_1(M) \cap h$ must be connected because the faces are convex. Note that this is not true if there are non-convex faces.

By 2.3 all the information about 0-tightness depends only on the 1-dimensional skeleton of the polyhedron. This 1-skeleton is nothing but a graph whose edges are straight line segments of Euclidean space, a so-called polyhedral graph. Therefore it is useful to characterize the 0-tightness for such graphs as follows:

2.4 Lemma: *An embedded and connected polyhedral graph $G \subset E^d$ is 0-tight if and only if the following conditions are satisfied:*

(i) G contains the 1-skeleton of its convex hull $\mathcal{H}G$,

(ii) every vertex of G which is not a vertex of $\mathcal{H}G$ lies in the relative interior of some of its neighbors.

PROOF: Let G be 0-tight. Then (i) follows from 2.2. Now let v be a vertex of G which is not a vertex of $\mathcal{H}G$. By the two-piece-property it is impossible to separate this vertex from its neighbors by a closed half space. This implies that the neighbors of v can never lie in an open half space, and (ii) follows.
To see the converse direction, let h be a closed half space such that $h \cap G$ is disconnected. One of these components certainly contains vertices of $\mathcal{H}$. If

there are several of those components then (i) is violated. Otherwise there is an 'interior component' of $h \cap G$. Move h orthogonal to its boundary until finally this interior component becomes as small as possible but still nonempty. It follows that this component contains a vertex which contradicts (ii).

EXAMPLES: Figure 1 gives three embeddings of the same graph G. In the second one G coincides with the 1-skeleton of its convex hull, the third one is not 0-tight according to 2.4.

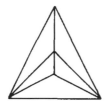  

Figure 1

For graphs in the Euclidean plane the 0-tightness just means that every component of the complement of the graph is either convex or the complement of a convex set. Therefore we obtain the following reformulation of a famous theorem in graph theory which in the original formulation characterizes the edge graphs of convex 3-polytopes:

THEOREM (E. Steinitz [Grü;13.1], also attributed to W.T. Tutte [Tu]): *Any 3-connected and planar graph admits a 0-tight embedding into the Euclidean plane.*

The theorem of Steinitz says that any such graph is the edge graph of a certain 3-polytope. The Schlegel diagram of this polytope gives a 0-tight embedding.

2.5 Lemma: *For a polyhedral surface $M \subset E^d$ the following conditions are equivalent:*

(i) M is 0-tight,

(ii) for every open or closed half space $h \subset E^d$ the induced homomorphism

$$H_*(h \cap M; \mathbf{Z}_2) \to H_*(M; \mathbf{Z}_2)$$

is injective where H_* denotes the standard singular homology.

In general, M is called tight if (ii) is satisfied, see 1.3. Lemma 2.5 just says that for compact surfaces without boundary tightness and 0-tightness are equivalent notions. The same holds for 1-dimensional complexes (graphs). In what follows we will often use this equivalence without further comment.

2.5 is a special case of a more general result in 3.18 below. We do not give a proof separately but just mention that it is a simple consequence of elementary Morse theory. (ii) $\Rightarrow$ (i) is obvious. Assume that (i) is true. For any linear function with finitely many critical points let μ_0, μ_1, μ_2 be the numbers of critical points of index 0, 1, 2, then we have $\mu_0 - \mu_1 + \mu_2 = \chi(M)$. Consequently, if $\mu_0 = \mu_2 = 1$ then $\mu_1 = 2 - \chi(M)$ is completely determined by the Euler characteristic of M. This implies that the inclusion $M \cap h \to M$ can never reduce the rank of the homology.

WARNING:
(a) 2.5 is not true for homology with arbitrary coefficients instead of $\mathbf{Z}_2$. If the surface is nonorientable then it holds for fields only of characteristic 2.
(b) 2.5 holds only for closed surfaces but not for surfaces with boundary. A cone over the boundary of a triangle is 0-tight but not tight.
(c) 2.5 does not hold for surfaces with singularities either, see the remark after 2.28.

2B. Tight surfaces with small codimension

In this section we give some basic results about the existence of tight polyhedral surfaces in the cases of small codimension, that are mainly the cases of codimension 1 and 2. Afterwards we turn to the question how large the essential codimension can be.

2.6 Theorem: *Given an abstract surface M with Euler characteristic $\chi(M)$, the following hold:*

(i) There is a tight polyhedral embedding $M \to E^3$ if M is orientable.

(ii) There is a tight polyhedral immersion $M \to E^3$ if M is nonorientable and $\chi(M) \leq -1$.

(iii) There are tight and substantial polyhedral embeddings $M \to E^4$ and $M \to E^5$ if M is distinct from the 2-sphere.

(iv) There is a tight and substantial polyhedral embedding $M \to E^6$ if M is orientable and distinct from the 2-sphere.

PROOF: The proof consists in a series of examples. It turns out that the most difficult case is the nonorientable surface in E^3 with $\chi = -1$, settled only recently by D. Cervone [Ce2]. The case $\chi = -3$ is also special although it is a consequence of the case $\chi = -1$ by attaching a handle. In fact, in this case there is a better solution which is fairly symmetric and, in addition, smoothable by a tight surface.

(i) The boundary of any convex 3-polytope provides a tight embedding $S^2 \to E^3$. In particular we can take the boundary of a cube (up to affine transformations). Then it is possible to cut g square-shaped holes into the top and the bottom of the cube and to join them by straight polyhedral cylinders, see Figure 2. It is easy to see that the resulting surface of genus g is tight. After a subdivision into convex polygons the conditions in 2.4 are satisfied by the edge graph. All these examples can easily be smoothed, still preserving the tightness.

A completely different example of a tight polyhedral torus in E^3 is given by Császár's torus with 7 vertices, based on the unique 7-vertex triangulation of the torus, see [Cs], [Bo-Eg]. This triangulation was already known to A. Möbius [Mö]. In this case the 1-skeleton is the complete graph K_7 with 7 vertices, see Figure 2. Therefore any simplexwise linear embedding is tight by 2.3 and 2.4. A polyhedral surface of genus $g \geq 1$ must have vertices at which the surface is not locally convex. For the necessary number of non-convex vertices for tight surfaces see [Gri]. An alternative construction of a tight orientable surface in E^3 was given in [Ba-Kui] as the boundary of the difference set of two convex polytopes which are dual to each other.

(ii) If $\chi(M) = -2$ we get an example by starting with a tight polyhedral torus constructed as in (i) (a cube with a rectangular hole). Then a non-orientable cylindrical handle can be attached, joining the outer part to the inner part, see Figure 2. This construction is due to N.H. Kuiper [Kui2]. As in (i) one can add arbitrarily many handles such that the tightness is preserved. This proves the assertion in the case of an even Euler characteristic. If $\chi(M)$ is odd we have to find a starting example. In [Kü-Pi] a tight polyhedral surface with $\chi = -3$ is given; the inner part, shown in Figure 2, is attached to an outer tetrahedron. The tightness follows from 2.3 and 2.4. Note that the curve of self-intersection with the triple point at its centre does not consist of edges. By attaching handles as above, the assertion follows for any $\chi \leq -2$. All these cases can be smoothed tightly, for an explicit smoothing procedure see [Kü-Pi].

The case $\chi = -1$ is very special because a theorem by F. Haab [Haa] says that there is no smooth tight immersion of this topological type. Supported by Haab's result, it had been conjectured that there is no polyhedral tight immersion either.

Surprisingly, this is not true, and in 1994 D. Cervone [Ce2] came up with an example. In Figure 2 we reproduce his triangulation. The vertices of the convex hull are a, c, d, e, f, g, h (b lies on the edge ac). Figure 2 shows the outer part (= a cylinder) and the inner part (= a Möbius band with hole) of the triangulation. The positions of the vertices are as follows:

$$
\begin{array}{llllll}
a & = & (\text{-}2,0,0), & e & = & (\text{-}2,\text{-}1,2), & i & = & (-\tfrac{3}{8},0,\tfrac{1}{2}) \\
b & = & (0,0,0), & f & = & (1,\text{-}1,2), & j & = & (\tfrac{1}{2},\tfrac{1}{4},1) \\
c & = & (1,0,0), & g & = & (1,1,2), & k & = & (-\tfrac{1}{4},\tfrac{7}{12},\tfrac{7}{6}) \\
d & = & (0,1,0), & h & = & (0,3,2), & l & = & (0,\tfrac{3}{4},\tfrac{7}{6}) \\
 & & & m & = & (\tfrac{1}{4},0,\tfrac{1}{2}).
\end{array}
$$

Again the tightness follows from the 0-tightness of the edge graph, 2.3 and 2.4.

(iii) A tight polyhedral torus in E^4 is given by the cartesian product of two planar convex polygons, e.g., by two squares. This is an analogue of the Clifford torus in S^3. The edge graph of this square torus coincides with the edge graph of a 4-dimensional cube. The tightness follows from 2.2. As in (i), it is possible to attach arbitrarily many orientable handles tightly. A nonorientable handle can be attached in the following way: In the 4-cube $C^4 = [0,1]^4$ take two opposite 2-faces, spanned by the vertices $(0,0,0,0)$, $(0,0,0,1)$, $(0,0,1,0)$, $(0,0,1,1)$, and the opposite face spanned by $(1,1,0,0)$, $(1,1,0,1)$, $(1,1,1,0)$, $(1,1,1,1)$. These two squares lie in a common 3-space. In this 3-space we can attach a cylindrical handle as in (i). The resulting surface is nonorientable and still tight. Its image under a suitable projection is coincides with the tight Klein bottle with handle in E^3 described above. Additional handles can be added. Note that all these examples can be smoothed tightly as in (ii). A tight Klein bottle in E^4 has been constructed by T. Banchoff [Ba5]: Start with two copies of a 5-vertex Möbius band (Figure 2) in two parallel hyperplanes of E^4 and join their boundaries by a straight cylinder.

This covers the case of even Euler characteristic in E^4. In the case of odd Euler characteristic we can start with a simplexwise embedding of the 6-vertex triangulation of the real projective plane (= half of the icosahedron), see Figure 2. The tightness of any simplexwise embedding is guaranteed by the completeness of the edge graph and by 2.3 and 2.4. Then handles can be attached tightly in any 3-space spanned by two adjacent triangles. The same argument shows that any surface with odd Euler charateristic admits a tight and substantial polyhedral embedding into E^5. In this case we start with the 6-vertex real projective plane, regarded as a subcomplex of the 5-dimensional simplex. The remaining case of surfaces with even Euler characteristic in E^5 can be treated in a similar way. We only need a starting example of a torus and of a Klein bottle . For the torus we can use a simplexwise linear embedding of the 7-vertex torus, see Figure 2, for the Klein bottle we can start with two 5-vertex Möbius bands in two parallel hyperplanes of E^5 and join their boundaries by a straight cylinder, a construction first given in [Ba5].

(iv) As in (iii), we start with a tight embedding of the torus: regard the 7-vertex triangulation as a subcomplex of the 6-dimensional simplex. This is tight by 2.3. Then handles can be added as described above.

This completes the proof of 2.6. The cases not covered by 2.6 are the following:

$M \to E^3$ for the real projective plane and for the Klein bottle. No tight immersion of this type exists, not even purely topologically, see [Kui10]. We do not give separate proofs for the polyhedral case.

$S^2 \to E^k$, $k \geq 4$, see 2.9 below,

$M \to E^k$, $k \geq 6$, see 2.14, 2.15 below.

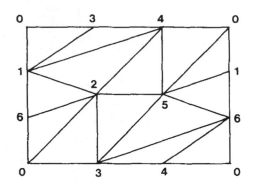 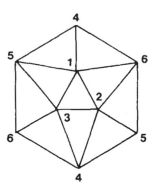

7-vertex torus   and 6-vertex real projective plane

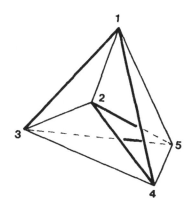

5-vertex Möbius band

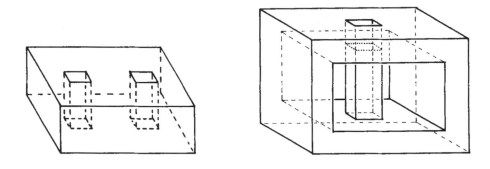

a tight orientable surface of genus 2   and a tight Klein bottle with one handle

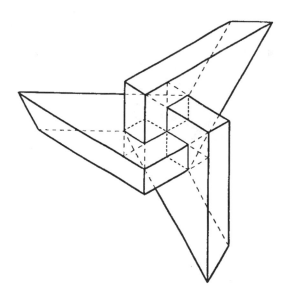

inner part of a tight surface with $\chi = -3$

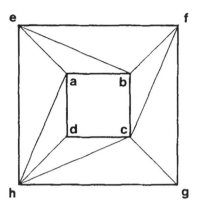

outer part of Cervone's surface with $\chi = -1$

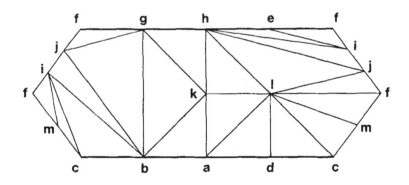

inner part of Cervone's surface with $\chi = -1$

Figure 2

2.7 Remark (Comparison with the smooth case): There is a perfect analogue of
2.6 (i), (ii) in the case of smooth tight immersions, except if $\chi(M) = -1$ [Haa],
[Ce2]. Tight smooth and substantial embeddings $M \to E^4$ exist whenever M is
orientable or nonorientable with $\chi(M) \le -2$. As mentioned in the proof of 2.6,
these can be constructed by smoothing tight polyhedral examples. The cases of
a tight Klein bottle in E^4 and a tight projective plane with one handle in E^4 are
still open. The only smooth and substantial tight surface in E^5 is the projective
plane, see [Kui3]. No tight smooth and substantial surfaces exist in E^k for $k \ge 6$.

U. Pinkall [Pi1] found smooth tight surfaces in E^3 in each regular homotopy class (with a few exceptions). These are essentially smoothings of tight polyhedral examples. More such polyhedral examples are due to D. Cervone [Ce1].

Recall from 2.2 that a 0-tight and connected polyhedron in Euclidean space has to contain the 1-skeleton of its convex hull. In order to obtain more information from this basic fact, the following lemma is very useful:

2.8 Lemma (B. Grünbaum): *The 1-skeleton of any convex d-polytope contains the complete graph K_{d+1} as a subset (not necessarily as a subgraph).*

PROOF: There is nothing to prove for $d = 1$, and the assertion is obvious for $d = 2$. For $d \geq 3$ we use induction on d. Let P be a d-polytope, then we pick a vertex v of P and a hyperplane H with $H \cap P = v$. A an inwards parallel displacement H' of H leads to a $(d-1)$-dimensional slice $P' = H' \cap P$ of P which is a convex $(d-1)$-polytope. By induction hypothesis the 1-skeleton of P' contains a K_d. This can be deformed into the 1-skeleton of P by central projection from v. Finally, P contains the cone over P' with apex v and therefore the cone over K_d which is nothing but a K_{d+1}.

2.9 Corollary [Ba4]: *Let $f : S^2 \rightarrow E^d$ be a tight and substantial polyhedral embedding. Then $d = 3$, and $f(S^2)$ is the boundary of a convex 3-polytope.*

PROOF: By 2.2 $f(S^2)$ contains the 1-skeleton of its convex hull. This in turn contains a K_{d+1} by 2.8. On the other hand S^2 does not contain a K_5 because K_5 is not planar, hence it does not contain a K_{d+1} for any $d \geq 4$. Therefore $d + 1 = 4$. Assume that $f(S^2)$ is not identical with its convex hull $\mathcal{H}$. Then one of the 2-dimensional faces of $\mathcal{H}$ is not contained in $f(S^2)$. On the other hand the boundary of this 2-face is certainly contained in $f(S^2)$ by 2.2. This boundary separates S^2 into two pieces. Therefore it follows that a certain plane in E^3, parallel to this 2-face, would dissect $f(S^2)$ into more than two pieces, a contradiction to the 0-tightness.

REMARK: A congruence theorem for tight polyhedral surfaces in E^3 does not hold in general: An example by T. Banchoff [Ba3] shows that there are tight polyhedral tori in E^3 which are non-congruent but intrinsically isometric. They even flex as metric spaces, being polyhedra at each intermediate stage. However, a conjecture of G. Kalai ([Kal1;13.23]) says that tight polyhedral surfaces in E^3 are rigid if the triangulation is fixed, i.e., there is no continuous deformation of the surface with the faces held rigid. By 2.9 this is true at least for genus zero because convex polyhedra are rigid [Con2]. Compare R. Connelly's flexing sphere [Con1] which is obviously not tight.

2.10 Lemma: *Let M be a connected polyhedron, embedded as a subcomplex of a convex d-polytope P such that $P = \mathcal{H}(M)$. Then M is 0-tight if and only if $Sk_1(P) \subset M$. In particular, if M is a surface, it is tight if and only if $Sk_1(P) \subset M$.*

PROOF: This is a direct consequence of 2.3 and 2.4 because by assumption all vertices of M are vertices of P. The necessity follows already from 2.2. To see the sufficiency we observe that for an arbitrary half space $h \subset E^d$ $h \cap Sk_1(P)$ is connected. Therefore $h \cap M$ is connected because each higher dimensional face is convex and connected to some edge.

2.11 Definition: A subcomplex of a convex polytope P is called 1-Hamiltonian if it contains $Sk_1(P)$. If P is a simplex we say also that a simplicial complex is 2-neighborly if it contains $Sk_1(P)$. 2-neighborliness of a simplicial complex means that any pair of vertices is joined by an edge in this complex. The notion of 1-Hamiltonian complexes generalizes the classical concept of Hamiltonian circuits containing $Sk_0(P)$.

2.10 says that a subcomplex $M \subset P$ is 0-tight if and only if it is 1-Hamiltonian. Banchoff's tight Klein bottle in E^5 is a 1-Hamiltonian subcomplex of the prism $\Delta^4 \times [0,1]$, for other examples of 1-Hamiltonian surfaces in prisms see [Sch1]. The 5-vertex Möbius band, the 6-vertex $\mathbf{R}P^2$, and the 7-vertex torus (Fig. 2) are 1-Hamiltonian in Δ^4, Δ^5, and Δ^6, respectively. As an illustration we mention the following examples.

2.12 Examples (Tight surfaces in the cube): For an arbitrary integer $d \geq 3$ one can construct a 1-Hamiltonian surface as a subcomplex $M(d)$ of the d-cube $C^d := [0,1]^d$ as follows: a 2-dimensional square Q is contained in $M(d)$ if and only if

$$Q = \{\epsilon_1\} \times \ldots \times \{\epsilon_{i-1}\} \times [0,1] \times [0,1] \times \{\epsilon_{i+2}\} \times \ldots \times \{\epsilon_d\}$$

for some index $i \in \{1, \ldots, d\}$ in cyclic order (i.e., $d + 1 := 1$), where $\epsilon_1, \ldots, \epsilon_d \in \{0,1\}$. The link of each vertex is a closed polygonal circuit of length d. Thus $M(d)$ is homeomorphic to a 2-manifold without boundary, and obviously every edge of C^d occurs in some $Q \subset M(d)$. In particular $M(3) = \partial(C^3)$ and $M(4) = \partial(C^2) \times \partial(C^2)$. By 2.10 $M(d)$ is tight and substantial in E^d. In short notation we may write $M(d) = 2^{\{d\}}$ where $\{d\}$ denotes the boundary of a regular d-gon in E^2, see 3.19. It is easily checked that $M(d)$ is an orientable surface of genus $g_d = 2^{d-3}(d-4) + 1$, the latter coincides with the genus of the d-cube (or rather its edge graph).

These examples have been given by T. Banchoff ([Ba1;Thm.A]) including a discussion of the tightness. As abstract 1-Hamiltonian subcomplexes of C^d

they have been independently observed before by H.S.M. Coxeter [Cox1], L.W. Beineke and F. Harary [BH], and G. Ringel [Ri2]. Nonorientable examples of the same type have been constructed by M. Jungerman [J] and by Ch. Schulz [Sch2]. For higher dimensional analogues see 4.2 and Section 3C.

As a consequence we see that the essential codimension of a tightly embedded polyhedral surface can be arbitrarily large. This raises the following question:

Depending on the genus of M, what is the maximal dimension d such that there exists a tight and substantial embedding $M \rightarrow E^d$?

2C. Higher codimension and the Heawood inequality

The answer to the question above is closely related with the existence of certain graph embeddings into given surfaces, a condition already mentioned in Lemma 2.2. More precisely, by Lemma 2.8 an essential necessary condition is the embeddability of the complete graph. This in turn is closely related with the Heawood map color problem, formulated by P.J. Heawood in 1890 and solved between 1950 and 1970.

2.13 Theorem (G. Ringel, J.W.T. Youngs, [Ri3]): *For every abstract surface M (except for the Klein bottle) the following conditions are equivalent:*

(i) There exists an embedding $K_n \rightarrow M$.

(ii) $\chi(M) \leq n(7 - n)/6$

(iii) $n \leq \dfrac{1}{2}(7 + \sqrt{49 - 24\chi(M)})$

(iv) $\dbinom{n - 3}{2} \leq 3(2 - \chi(M)) = 6g \quad (g = genus\ of\ M).$

For the Klein bottle (i) is equivalent to $n \leq 6$. Moreover, equality in the inequalities implies that the embedding of K_n induces an abstract triangulation of M.

(ii), (iii), and (iv) are known as *Heawood's inequality*, compare [Hea], [Wh1].

PROOF:

(i) $\Rightarrow$ (ii): We start with $K_n \subset M$ $(n \geq 3)$ and observe that each component C of $M \setminus K_n$ is an open surface with $\chi(C) \leq 1$ which is bounded in M by a certain number of edges of K_n. Let c_i denote the number of such components bounded by $i \geq 3$ edges. This implies

$$2\binom{n}{2} = \sum_{i \geq 3} i \cdot c_i \geq 3 \cdot \sum_i c_i$$

and furthermore

$$\chi(M) = n - \binom{n}{2} + \chi(M \setminus K_n)$$

$$\leq n - \binom{n}{2} + \sum_i c_i$$

$$\leq n - \binom{n}{2} + \frac{2}{3}\binom{n}{2}$$

$$= n(7 - n)/6$$

with equality if and only if each component of $M \setminus K_n$ is an open disc, bounded by three edges of K_n, i.e., if and only if the embedding of K_n induces an abstract triangulation of M.

(ii) $\Leftrightarrow$ (iii) $\Leftrightarrow$ (iv) holds by a simple algebraic transformation.

(iii) $\Rightarrow$ (i) is the main part of the Map Color Theorem. The proof covers the whole book [Ri3], compare [Wh1]. It is given separately for each case $n \equiv 0, 1, 2, \ldots, 11 \bmod 12$ if M is orientable and $n \equiv 0, 1, 2 \bmod 3$ if M is nonorientable. For small values of n several exceptional cases have to be considered. The relationship with the coloring problem is the following: regarding the vertices of a graph as countries, an edge between two vertices corresponds to a common border of the countries. This is a duality principle. It is fairly clear that the dual of a $K_n \subset M$ gives a map on M which requires at least n distinct colors for coloring. With respect to the result, the only exception is the case of the Klein bottle which does not admit an embedding of K_7 although (iii) is satisfied for $n = 7$. This particular case is due to P. Franklin [Fr]. The following Figure 3 shows a triangulation of the Klein bottle which contains a K_6 in its edge graph. For a triangulation of the surface with $\chi = -1$ containing a K_7 see Figure 4.

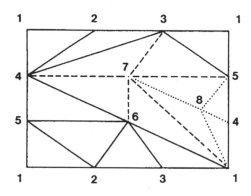

Figure 3

2.14 Theorem (W. Kühnel, [Kü2]): *Let M be an abstract surface and let $n \geq 6$ be a given number. Then the following conditions are equivalent:*

(i) There exists a tight and substantial polyhedral embedding $M \to E^{n-1}$.

(ii) There exists an embedding $K_n \to M$.

The same holds for surfaces with boundary if we replace 'tight' by '0-tight'.

PROOF:

(i) $\Rightarrow$ (ii) : Let $\mathcal{H}$ denote the convex hull of M in E^{n-1}. The 1-skeleton of $\mathcal{H}$ is contained in M by 2.2, and the complete graph K_n is contained in $Sk_1(\mathcal{H})$:

$$K_n \subset Sk_1(\mathcal{H}) \subset M \subset E^{n-1}.$$

(ii) $\Rightarrow$ (i) : We start with an embedding $K_m \subset M$ where $m \leq n$ is maximal with respect to the inequality 2.13 (iii). Then we extend it to a triangulation of M with those m vertices and some extra vertices. Finally the m vertices can be put into general position in E^{n-1}, and the extra vertices have to be chosen in the relative interiors of their neighbors, according to 2.4. This implies that the edge graph of this triangulation is 0-tight in E^{n-1}. If the surface is embedded (i.e., without self-intersections) then it is 0-tight and tight by 2.3, 2.4, and 2.5. The only difficulty arises if two extra vertices have exactly the same four neighbors. This would lead to self-intersections in the interior of the tetrahedron spanned by those four vertices. However, this situation can be avoided by changing the triangulation. The details involving many results from [Ri2] are given in our previous paper [Kü2].

An example is the case of a triangulated surface with $\chi = -1$ containing a K_7, see Figure 4.

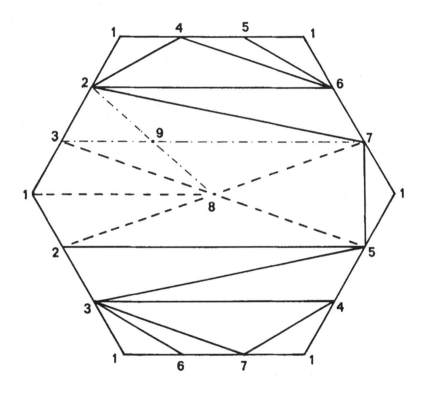

Figure 4

2.15 Corollary ([Kü2;Thm.A]): *Let M be an abstract surface which is distinct from the Klein bottle . Then there exists a tight substantial polyhedral embedding $M \to E^d$ if and only if*

  (i) $d \geq 3$ *(if M is orientable)* or $d \geq 4$ *(if M is nonorientable)*, *and*

  (ii) $d \leq \frac{1}{2}(5 + \sqrt{49 - 24\chi(M)})$.

The Klein bottle can be embedded tightly and substantially into E^4 and E^5 but not into E^k, $k \geq 6$.

This follows from 2.13 and 2.14 in connection with 2.6. The upper bound for d is originally due to T. Banchoff [Ba1;p.470]. However, Theorem B in [Ba1] gives the non-equivalent formulation $d < \frac{1}{2}(7 + \sqrt{49 - 24\chi(M)})$ which is not sharp if the right hand side is not an integer. For the case of the Klein bottle compare [Ba5;Thm.K] and 2.6 (iii). In Section 2 of [Ba5] two different tight embeddings of the Klein bottle into E^5 are given. A third type of such an embedding uses the triangulation shown in Figure 3. By the construction principle mentioned in the proof of 2.14 we put the vertices $1, 2, \ldots, 6$ in general position in E^5, 7 in the relative interior of $\langle 1, 3, 4, 5, 6 \rangle$, and 8 in the relative interior of $\langle 1, 4, 5, 7 \rangle$.

2D. Tight triangulations of surfaces

A triangulated surface (or any simplicial complex) with n vertices can always be regarded as a subcomplex of an $(n-1)$-dimensional simplex $\triangle^{n-1}$. We call this the canonical embedding of the triangulation. An abstract triangulation of a surface is called tight if its canonical embedding is tight (compare 3.2 and 3.17).

2.16 Corollary (Tight triangulations): *Let M be a triangulated surface with n vertices. Then the following conditions are equivalent:*

(i) The triangulation is tight.

(ii) The triangulation is 2-neighborly, i.e., its edge graph is a complete graph K_n.

(iii) $\chi(M) = n(7 - n)/6$

(iv) $n = \dfrac{1}{2}(7 + \sqrt{49 - 24\chi(M)})$

(v) $\dbinom{n-3}{2} = 3(2 - \chi(M)) = 6g$ (g = genus of M).

Conversely, given an abstract surface (which is distinct from the Klein bottle) and a number n satisfying (iii) or (iv) or (v), then there is a tight triangulation of M with n vertices.

PROOF: Let f_1 and f_2 denote the number of edges and triangles, respectively.

(i) $\Leftrightarrow$ (ii) follows from 2.10 and 2.11 if we consider the canonical embedding $M \to \triangle^{n-1}$.

(ii) $\Rightarrow$ (iii): $\chi(M) = n - f_1 + f_2 = n - \frac{1}{3}f_1 = n - \frac{1}{3}\binom{n}{2} = n(7-n)/6$.

(iii) $\Rightarrow$ (ii): $n - \frac{1}{3}\binom{n}{2} = \frac{1}{6}n(7-n) = \chi(M) = n - \frac{1}{3}f_1$.

(iii) $\Leftrightarrow$ (iv) $\Leftrightarrow$ (v) is a simple algebraic computation, as in 2.13.

The converse assertion for an abstract surface M follows from 2.13.

REMARK (Algebraic curves): A nonsingular curve of algebraic degree d in the complex projective plane is an orientable surface of genus $g = \binom{d-1}{2}$. By 2.16 tight triangulations of such curves with n vertices exist if $\binom{n-3}{2} = 6\binom{d-1}{2}$. There is an infinite sequence of such pairs (d_i, n_i) recursively defined by

$$d_1 = 1, \; d_2 = 3, \; d_3 = 5, \; d_4 = 16, \quad d_{i+2} = 10d_i - d_{i-2} - 13,$$

$$n_1 = 4, \; n_2 = 7, \; n_3 = 12, \; n_4 = 39, \quad n_{i+2} = 10n_i - n_{i-2} - 28.$$

2.17 Theorem (T. Banchoff, W. Pohl, [Ba5], [Po]): *Let $M \subset E^d$ be a tightly and substantially embedded polyhedral surface. Then the following hold:*

(i) $\displaystyle \binom{d-2}{2} \leq 3(2 - \chi(M)).$

(ii) *If $d \geq 4$ then equality in (i) holds if and only if M is embedded as a subcomplex of a d-simplex Δ^d (and the induced triangulation is tight by 2.16).*

Equality in (i) is satisfied by the boundary of any convex 3-polytope. This shows that (ii) cannot be extended to the case $d = 3$.

PROOF: (i) As mentioned before, 2.3 and 2.4 together lead to an embedding $K_{d+1} \to M$, and the inequality is just Heawood's inequality from 2.13.
(ii) We start with $K_{d+1} \subset Sk_1(\mathcal{H}) \subset M$. By the proof of (i) $\Rightarrow$ (ii) in 2.13 we know that each component of $M \setminus K_{d+1}$ is topologically an open disc. Therefore each component of $M \setminus Sk_1(\mathcal{H})$ is also an open disc, bounded by a certain number of edges of $\mathcal{H}$. Let d_i denote the number of such components which are bounded by $i \geq 3$ edges of $\mathcal{H}$. Let n be the number of vertices and f_1 the number of edges of $\mathcal{H}$.

This implies

$$2f_1 \;\; = \textstyle\sum_{i \geq 3} i\, d_i \geq 3 \sum_i d_i$$

$$2f_1 \;\; \geq n \cdot d$$

$$n \;\; \geq d + 1$$

and, consequently,

$$\begin{aligned}
\tfrac{1}{6}(d+1)(6-d) &= \chi(M)\\
&= n - f_1 + \sum_i d_i\\
&\leq n - f_1 + \tfrac{2}{3}f_1\\
&\leq n - \tfrac{1}{6}n \cdot d\\
&= \tfrac{1}{6}n(6-d).
\end{aligned}$$

At this point we have to distinguish between four cases:

$d \geq 7$ immediately implies $n = d+1$ which means that $\mathcal{H}$ is a d-simplex Δ^d and $K_n = Sk_1(\mathcal{H})$. By the tightness it follows that each abstract 3-gon in $M \setminus K_n$ is embedded as a planar triangle, which then is part of $Sk_2(\Delta^d)$. Consequently, that M is a subcomplex of $Sk_2(\Delta^d)$, compare [Ba5;Thm.M].

$d = 4$ is not a case of equality in (i).

The remaining cases $d = 5$ and $d = 6$ are more complicated. Let us first recall the *combinatorial Gauss-Bonnet formula*. For a polyhedral surface M with vertices $v_1, \ldots, v_m$ let $K_1, \ldots, K_m$ be the polyhedral curvatures at these vertices. K_i is defined as 2π minus the sum of the interior angles of all the faces meeting at v_i. Then the combinatorial Gauss-Bonnet formula states $\sum_i K_i = 2\pi\chi(M)$, compare [Ba2], [Br-Kü1]. In particular, if e_i denotes the number of edges emanating from v_i, then the mean value of $6 - e_1, \ldots, 6 - e_m$ has the same sign as $\chi(M)$. For a tight polyhedral surface, 2.4 implies the following: a vertex v which is not a vertex of the convex hull satisfies $K(v) \leq 0$ where $K(v) = 0$ can occur only if v has exactly three neighbors. In the latter case v is an interior vertex of the planar triangle spanned by its three neighbors; we therefore do not count v as a proper vertex. The proper vertices of M which are not vertices of the convex hull have strictly negative polyhedral curvature. Conversely, a vertex with strictly positive curvature must be a vertex of the convex hull.

If $d = 5$ then $\chi(M) = 1$, i.e., M is the real projective plane. This case is thoroughly treated by T. Banchoff in [Ba5], and we repeat his arguments here. By the combinatorial Gauss-Bonnet formula there is at least one vertex v from which exactly five edges $vw_1, \ldots, vw_5$ emanate, each being an edge of $\mathcal{H}$. Each of these edges vw_i is contained in at most two 2-faces of M but in at least four 2-faces of $\mathcal{H}$. This means that there are at least two essential (i.e., non-contractible) 2-top-sets containing vw_i. The boundary of any essential 2-top-set (a *top-polygon*) is a convex polyhedral curve homologous to a projective line in $\mathbf{R}P^2$. From the intersection form of $\mathbf{R}P^2$ we know that any two of those must intersect. Without loss of generality we can choose the numbering so that vw_i and vw_{i+2} determine an essential 2-top-set for $i = 1, \ldots, 5$ (cyclic labelling).

We want to show that all 2-faces of $\mathcal{H}$ which are incident with v are triangles. Assume that vw_1, vw_2 and w_1u are contained in a 2-face and that $u \neq w_2$. There is an essential top-polygon T at the edge w_1u which does not meet vw_2. This has to meet the top-polygon determined by vw_2 and vw_4 at a point not on vw_2. Therefore T is contained in the 3-top-set determined by v, w_1, w_2, w_4. The plane determined by v, w_3, w_5 can meet this 3-top-set only at the point v which is not in T. Therefore T does not meet the top-polygon determined by vw_3 and vw_5, a contradiction.

Similarly, if vw_1, vw_3 and v_1u ($u \neq w_3$) are contained in a 2-face then we proceed in the same way, interchanging the roles of the other vertices. In any case we get a contradiction. This implies that such an u cannot occur and that all the 2-faces of $\mathcal{H}$ at v are triangles.

Next we show that five of them must be contained in M. Assume that the triangle vw_1w_2 is a face of $\mathcal{H}$ with an interior vertex z such that w_1w_2z is not contained in M but the quadrilateral vw_1zw_2 is. Then w_1w_2z is essential, i.e., not homologous to zero in M. Therefore it has to meet the top-polygon determined by vw_3w_5, a contradiction. For vw_2w_3, vw_3w_4 etc. the same argument applies.

Now we know that at v there meet exactly five triangles of M and exactly five essential top-polygons which are also triangles. The question is: Are there vertices u of $\mathcal{H}$ distinct from $v, w_1, \ldots, w_5$? Assume this is true, and assume that w_1u is an edge of $\mathcal{H}$. Any essential top-polygon T containing w_1u would then have to meet the top-polygons determined by vw_2w_4, vw_2w_5 and vw_3w_5. These intersections must be vertices of $\mathcal{H}$. This is impossible because v is not in T and because all these are distinct 2-faces of $\mathcal{H}$.

It follows that $\mathcal{H}$ has only six vertices ($\mathcal{H} = \Delta^5$), and that M is contained in $Sk_2(\mathcal{H})$.

If $d = 6$ then $\chi(M) = 0$, i.e., M is the torus since the Klein bottle is excluded by 2.13. In this case the inequality above reads as

$$0 = \chi(M) \leq n - \tfrac{1}{3}f_1 \leq n - \tfrac{1}{6}nd = 0$$

which in particular implies $2f_1 = 6n$. Therefore $\mathcal{H}$ is a simple 6-polytope (Def. 1.1), each of its k-dimensional faces being a simple k-polytope. At each vertex of $\mathcal{H}$ there meet 6 edges, each being contained in M. It follows that there must be at least six 2-faces of M around each vertex of $\mathcal{H}$. By the combinatorial Gauss-Bonnet formula for M the mean value of the polyhedral curvatures of M at the vertices of $\mathcal{H}$ is nonpositive, and it is strictly negative unless there are exactly six 2-faces of M around each vertex of $\mathcal{H}$. The proper vertices of M which are not vertices of $\mathcal{H}$ have strictly negative curvature. Therefore no such vertex can occur. It follows that all 2-faces of M are also 2-faces of $\mathcal{H}$. Furthermore, every 2-face of M must be a triangle because otherwise the sum of the interior angles would lead to a contradiction to the combinatorial Gauss-Bonnet formula.

At this point it follows that the universal covering of this torus is combinatorially equivalent to the classical tessellation {3,6} of the plane by regular triangles. However, it is not yet clear what the number of vertices of M is.

Let A be a 3-face of $\mathcal{H}$. By the tightness and by $rk(H_1(M)) = 2$ the 3- top-set $A \cap M$ is homeomorphic to a 2-sphere with at most three holes. Since all 2-faces of M are triangles, it follows that ∂A has at most three k-gonal faces, $k \geq 4$. If there is an i-gon in A with $i \geq 6$ then there must be at least three other j-gons with $j \geq 4$ since any two triangles must be disjoint. Here we use the fact that A is a simple 3-polytope. It follows that ∂A consists of p_3 triangles, p_4 quadrilaterals and p_5 pentagons where $p_4 + p_5 \leq 3$.

By V. Eberhard's theorem (see [Grü] 13.3, in particular Table 13.3.1) A must be either a tetrahedron or a triangular prism. If A is a tetrahedron then $A \cap M$ contains one or two triangles of M (and two or three empty triangles), if A is a prism then $A \cap M$ contains the two triangles.

Now let B be a 4-face of $\mathcal{H}$. If ∂B contains a tetrahedron then it must contain another one adjacent to it along an empty triangle of M. Since B is simple, it must be a simplex. Otherwise each facet of B is a triangular prism. But in this case the triangles in $B \cap M$ are isolated and cannot form any nontrivial 2-chain in simplicial homology. Therefore the rank of $H_1(B \cap M)$ is greater than 2 contradicting the tightness. Therefore this case does not occur, and B is a simplex. It follows that each 5-face of $\mathcal{H}$ is simplicial and simple, hence a simplex. Finally, the same argument implies that $\mathcal{H}$ itself is a simplex.

This completes the proof of (ii). The structure of a different proof for the case $d = 6$ in (ii) can be found in [Po;§5]. The same conclusion for tight topological immersions of the torus into E^6 is claimed in [Po; §1,Thm.5], compare 2.20.

There is a much shorter proof for (ii) under the additional assumption that $\mathcal{H}$ is 2-simplicial, i.e., if every 2-face is a simplex. In this case G. Kalai's improvement of the Lower Bound Theorem [Kal1;3.18] states $f_1 \geq n \cdot d - \binom{d+1}{2}$. As in the beginning of the proof of (ii) we conclude:

$$
\begin{aligned}
\tfrac{1}{6}(d+1)(6-d) &= \chi(M) \\[2mm]
&\leq n - \tfrac{1}{3}f_1 \\[2mm]
&\leq n - \tfrac{1}{3}\left(n \cdot d - \binom{d+1}{2}\right) \\[2mm]
&= \tfrac{1}{3}n(3-d) + \tfrac{1}{3}\binom{d+1}{2} \\[2mm]
&\leq \tfrac{1}{3}(d+1)(3-d) + \tfrac{1}{3}\binom{d+1}{2} \\[2mm]
&= \tfrac{1}{6}(d+1)(6-d)
\end{aligned}
$$

which implies $n = d + 1$ whenever $d \geq 4$. Evidently no conclusion about n is possible if $d = 3$.

2.18 Corollary ([Kü7]): *Let P be a convex d-polytope containing a 1-Hamiltonian surface M in its 2-skeleton. Then the following hold:*

(i) $\dbinom{d-2}{2} \leq 3\,(2 - \chi(M))$.

(ii) *If $d \geq 4$ then equality in (i) holds if and only if P is a simplex.*

2.19 Corollary ([Ba5], [Po]): *The image of a tight polyhedral real projective plane in E^5 or of a tight polyhedral torus in E^6 is unique up to affine transformations. It is the image of the canonical embedding of the 6-vertex $\mathbf{R}P^2$ or of the 7-vertex torus (shown in Figure 2)*

The proof of 2.19 follows from 2.17 in connection with the well known fact that the 6-vertex triangulation of the real projective plane and the 7-vertex triangulation of the torus are combinatorially unique.

REMARK: It is surprising that in the cases of 2.19 the tightness (or TPP) puts extremely strong restrictions on the polyhedron and leads to a unique and moreover very regular object. In fact, the 6-vertex $\mathbf{R}P^2$ and the 7-vertex torus are twisted regular honeycombs in the sense of H.S.M. Coxeter: in his notation, $\{3,5\}_5$ and $\{3,6\}_{1,2}$, see [Cox2]. For the canonical embedding into a regular simplex $\triangle^5$ or $\triangle^6$ the full automorphism group of the triangulation (of order 60 or 42, respectively) is contained in the Euclidean group.
A uniqueness result as in 2.19 cannot be expected for any other topological type of a surface. For the canonical embeddings the geometrical classification (up to affine transformations) coincides with the combinatorial classification (up to permutations of the vertices). On the other hand there are two combinatorially distinct tight 9-vertex triangulations of the surface with $\chi = -3$, and there are 14 combinatorial types of tight 10-vertex triangulations of the surface with $\chi = -5$, see [AB], the latter ones are exactly the items $1 - 14$ in the list of block designs in [CCHR]. Furthermore A. Altshuler found 59 distinct 12-vertex triangulations of the orientable surface with $\chi = -10$ (see [ABS2]) and more than 40000 different ones of the corresponding nonorientable surface.

2.20 Conjecture (W. Pohl, [Po]): *Let* $M \to E^d$ *be a tight and substantial topological immersion of a surface, and assume* $d \geq 6$. *Then the following hold:*

(i) *The convex hull of* M *in* E^d *is a convex polytope.*

(ii) $\binom{d-2}{2} \leq 3(2 - \chi(M))$.

(iii) *Equality in (ii) holds if and only if* M *is embedded as a subcomplex of a* d-*simplex* Δ^d *(and the induced triangulation is tight by 2.16).*

For $d \leq 5$ part (i) is certainly not true because the Veronese surface is tight but its convex hull is not polyhedral. A theorem of Kuiper and Pohl [Kui-P] states that the Veronese surface and the 6-vertex triangulation as a subcomplex of a 5-simplex are the only image sets of tight topological embeddings of $\mathbf{R}P^2$ into E^5, up to projective transformations of the ambient space. If (i) is true then (ii) and (iii) follow by the same arguments as in the proof of 2.17. Pohl's proof of (i) seems to be lost. We therefore call it *Pohl's conjecture*.

The case of equality in the Heawood inequality (see 2.13, 2.17) coincides with the case of tight triangulations by 2.16. It is also emphasized by the following reversed inequality which comes in if one asks for triangulations with the minimal possible number of vertices. Let us repeat that triangulations are always understood as simplicial complexes, i.e., between two vertices there is at most one edge, and three vertices determine at most one triangle.

2.21 Theorem (Minimal triangulations of surfaces, G. Ringel, M. Jungerman, [Ri1], [J-Ri]): *Let* M *be an abstract surface which is distinct from the Klein bottle, the orientable surface of genus 2, and from the surface with* $\chi = -1$. *Then the following conditions are equivalent:*

(i) *There exists a triangulation of* M *with* n *vertices.*

(ii) $\binom{n-3}{2} \geq 3(2 - \chi(M))$.

Equality in (ii) holds if and only if the triangulation is tight.

The proof of (i) $\Rightarrow$ (ii) is trivial: $\chi(M) = n - \frac{1}{3}f_1 \geq n - \frac{1}{3}\binom{n}{2} = n(7-n)/6$. For the tightness apply 2.16. The hard part (ii) $\Rightarrow$ (i) was shown by G. Ringel [Ri1] in the nonorientable case and by M. Jungerman and G. Ringel [J-Ri] in the orientable case. They constructed a triangulation of M with the smallest number n of vertices satisfying (ii). The results and the proof methods of the Map Color Theorem [Ri3] are heavily involved. In the three exceptional cases the left hand side of (ii) has to be replaced by $\binom{n-4}{2}$, compare [Hun].

2.22 Theorem (Centrally-symmetric tight surfaces): *Assume that $M \subset E^d$ is a tight polyhedral surface whose convex hull is a centrally-symmetric simplicial d-polytope. Then the following hold:*

(i) $2(d-1)(d-3) \leq 3\bigl(2 - \chi(M)\bigr)$.

(ii) *For $d \geq 4$ equality in (i) holds if and only if - up to affine transformations - M is a subcomplex of the d-dimensional cross-polytope with $n = 2d$ vertices (the dual of the d-cube, [Cox-M]).*

(iii) *For $d \geq 5$ equality in (i) implies that the induced triangulation of M has the minimum number $n = 2d$ of vertices according to 2.21.*

(ii) does not hold for $d = 3$, and (iii) does not hold for $d = 3, 4$.

PROOF: (i) By assumption the convex hull $\mathcal{H} = \mathcal{H}(M)$ is a centrally-symmetric simplicial polytope with $f_0 \geq 2d$ vertices. For the number f_1 of edges the following inequality is known as part of R. Stanley's Lower Bound Theorem [Sta4]:

$$f_1 \geq d(f_0 - 2) \geq 4\binom{d}{2}.$$

By the tightness we have $Sk_1(\mathcal{H}) \subset M$ according to 2.2.
As in the proof of 2.13 (i) we conclude

$$
\begin{aligned}
\chi(M) &= f_0 - f_1 + \chi(M \setminus Sk_1(\mathcal{H})) \\
&\leq f_0 - f_1 + \sum_{i \geq 3} c_i \\
&\leq f_0 - f_1 + \tfrac{2}{3}f_1 \\
&= f_0 - \tfrac{1}{3}f_1 \\
&\leq f_0 - \tfrac{1}{3}d(f_0 - 2) \\
&= f_0(1 - \tfrac{d}{3}) + \tfrac{2}{3}d \\
&\leq 2d(1 - \tfrac{d}{3}) + \tfrac{2}{3}d \\
&= -\tfrac{2}{3}(d-1)(d-3) + 2.
\end{aligned}
$$

(ii) If $d \geq 4$ then equality in (i) implies $f_0 = 2d$, $f_1 = d(f_0 - 2) = 4\binom{d}{2}$ and that each component of $M \setminus Sk_1(\mathcal{H})$ is homeomorphic to a disc, bounded by three edges of $Sk_1(\mathcal{H})$.

An elementary argument shows that any centrally-symmetric d-polytope with $2d$ vertices is affinely equivalent to the d-dimensional cross-polytope because it is the convex hull of two antipodal facets with d vertices each. By the tightness each component of $M \setminus Sk_1(\mathcal{H})$ is a planar triangle. Thus M is a subcomplex of the 2-skeleton of its convex hull.

For $d = 3$ the boundary of any centrally-symmetric simplicial polytope is an example, e.g., the icosahedron.

(iii) Equality in (i) implies that M is triangulated with $n = 2d$ vertices and $f_1 = 4\binom{d}{2}$ edges. Observe first that there are d missing edges (= the diagonals) in M by the equation $\binom{2d}{2} = f_1 + d$.
Let us verify the inequality in 2.21:

$$\binom{2d-3}{2} - 3(2 - \chi(M)) = \binom{2d-3}{2} - 2(d-1)(d-3)$$
$$= 2d^2 - 7d + 6 - 2d^2 + 8d - 6$$
$$= d$$
$$> 0.$$

Now let us check whether a triangulation with $n - 1 = 2d - 1$ vertices is possible:

$$\binom{2d-4}{2} - 3(2 - \chi(M)) = \binom{2d-4}{2} - 2(d-1)(d-3)$$
$$= 2d^2 - 9d + 10 - 2d^2 + 8d - 6$$
$$= 4 - d$$

which is negative for $d \geq 5$, in disagreement with 2.21. This proves (iii).
In the exceptional case $d = 4$ and $\chi = 0$ the Klein bottle cannot be a subcomplex of the boundary complex of any 4-polytope, and an 8-vertex triangulation of the torus is not minimal because there is a 7-vertex triangulation. An example is the following: Regard the set of vertices as $\mathbb{Z}_8$ and take the $\mathbb{Z}_8$-orbits of the triangles $\langle 124 \rangle$ and $\langle 134 \rangle$. This can be regarded as a subcomplex of the 4-dimensional cross-polytope with diagonals $\langle 15 \rangle, \langle 26 \rangle, \langle 37 \rangle, \langle 48 \rangle$. It contains all $\binom{8}{2} - 4 = 24$ edges of the cross-polytope and is therefore tight by 2.10. Equality in (i) is satisfied. Infinitely many examples for the case of equality in (i) can be found in [J-Ri1].

CONJECTURE: *If $M \to E^d$ is any substantial tight polyhedral torus which is centrally-symmetric, then $d \leq 4$.*

Even if this conjecture is true, a geometrical uniqueness result cannot be expected for $d = 4$ because besides the example above there is another tight torus as a subcomplex of the 4-cube which is also centrally-symmetric, see 2.12.

2E. Surfaces with boundary

In this section we briefly discuss the case of tight polyhedral surfaces with boundary. In this case tightness and 0-tightness are inequivalent conditions. Actually the tightness is somewhat stronger. To see the difference consider a convex ball in E^n and a hemisphere in E^{n+1} (or polyhedral versions of those).

For the relationship between tightness and 0-tightness, one basic observation is the following lemma:

2.23 Lemma ([Kü2; Lemma 7]): *Let $M \subset E^d$ be a tight polyhedral surface with boundary $\partial M \neq \emptyset$. Then (i) and (ii) are equivalent:*

(i) *M is tight (with respect to any field).*

(ii) *M is 0-tight, and ∂M contains each vertex of $\mathcal{H}(M)$.*

In particular, if there is a tight polyhedral embedding of a surface M with one boundary component into E^d then there is a tight embedding of the closed surface of the same genus into E^{d+1} because we can close it up by the cone over ∂M, see [Kü2;Lemma 9]. It immediately follows from 2.19 that a tight polyhedral Möbius band in E^4 or a tight polyhedral torus with hole in E^5 is unique. From the remark after 2.19 we see that no such uniqueness result can hold for any other type of a closed surface with one hole. For the case of higher dimension compare 6.1 and 6.2.

PROOF: (i) $\Rightarrow$ (ii) We have to show that every vertex of $\mathcal{H}(\mathcal{M})$ lies on ∂M: Assume there is a vertex v of $\mathcal{H}(\mathcal{M})$ in $M \setminus \partial M$. Let h be a half space with $h \cap M = v$. Take another half space h' whose boundary is parallel to that of h and such that $h \cap h' = \emptyset$. Consider the inclusion $M \cap h' \to M$. $M \cap h'$ is homeomorphic to M minus an open disc. Therefore the rank of the first homology of $M \cap h'$ is greater than the rank of the first homology of M. This contradicts the tightness.

(ii) $\Rightarrow$ (i) In terms of height functions with finitely many critical points we have $\mu_0 - \mu_1 + \mu_2 = \chi(M) = 1 - \dim H_1(M; \mathbf{Z}_2)$, compare the proof of 2.5.

The equation $\mu_0 = 1$ expresses the 0-tightness. Therefore the tightness is equivalent to $\mu_2 = 0$. This in turn is guaranteed by the condition that on the boundary of $\mathcal{H}(M)$ there is no maximum which lies in $M \setminus \partial M$. A formal definition of μ_i (in more generality) will be given in Section 3B below.

For a compact surface M without boundary let M_r denote M with $r \geq 1$ separated open discs removed.

There is the following analogue of 2.6 for surfaces with boundary:

2.24 Theorem:

(i) If $M \cong S^2$ then there is a tight polyhedral embedding of M_r into E^2. If M is orientable of genus $g \geq 1$, then there is a tight polyhedral immersion of M_r into E^2 provided that $r \geq 2$.

(ii) For each surface M_r (except for the disc) there is a tight polyhedral embedding into E^3.

(iii) For each surface M_r (except if $M \cong S^2$) there is a tight and substantial polyhedral embedding into E^4, and there is one into E^5 except if $r \leq 2$ and if M is the projective plane or if $r = 1$ and if M is the Klein bottle .

PROOF: As a general observation we mention the following direct consequence of 2.23: If we start with a tight surface with nonempty boundary, then cutting a convex hole out of the interior of any of its 2-faces leads again to a tight surface with one more boundary component.

(i) If $M \cong S^2$ then every convex polygon with $r - 1$ convex holes provides an example. If M is a torus and $r = 2$, an immersion into E^2 is shown in Figure 5. For higher genus a similar construction is possible. By cutting out extra holes one can get an arbitrary $r \geq 2$. For tight immersions of surfaces into E^2 compare [Ro].

(ii) If $M \cong S^2$ and $r = 2$ an example is provided by an ordinary cylinder $\partial \Delta^2 \times [0,1]$. If $M \cong \mathbf{R}P^2$ and $r = 1$ we take a 5-vertex Möbius band embedded into E^3 (compare [Kui6]). If M is the torus or the Klein bottle and $r = 1$, we can take the examples shown in the second row of Figure 5 which were given by L. Rodriguez [Ro]. For higher values of r one has to cut out additional holes. For higher genus one can attach handles as in the case of closed surfaces in the proof of 2.6.

(iii) We start with the torus with one hole in E^5. In this case we cut out the open star of one vertex of the 7-vertex triangulation (Figure 2). Then we put the 6 vertices in general position in E^5. This surface is also a 5-top-set of the tight torus in E^6, and is therefore tight (1.4). It is also tight by Lemma 2.23. A projection into E^4 will also be tight. In fact, this 6-vertex torus-with-hole is a tight triangulation. Then one can cut out additional holes and attach handles. This covers the orientable case.

The same construction leads to a tight projective plane with a hole (= Möbius band) in E^4. The tight 5-vertex triangulation of the Möbius band has already been mentioned in the proof of 2.6. Then handles and crosscaps can be attached according to [Kü2;Lemma 6]. We can avoid the crosscaps by starting with the tight Klein bottle in E^5 from 2.15 which uses the triangulation shown in Figure

3. The procedure of cutting out the open star of the vertex 2 corresponds to taking the 4-top-set opposite to the vertex 2. This is a tight Klein bottle with one hole in E^4. It remains to attach handles and to cut out additional holes. This completes the nonorientable case in E^4.

In order to obtain a tight projective plane with 3 holes in E^5 we start with the canonical embedding of the 6-vertex triangulation. Then for each of the three edges $\langle 14 \rangle, \langle 25 \rangle, \langle 36 \rangle$ we cut out a small triangle with this edge lying on the boundary, see Figure 5. The resulting surface is tight in E^5 by 2.23.

Similarly for the Klein bottle with 2 holes we take the triangulation shown as the last part of Figure 5. This induces a tight embedding of the Klein bottle into E^5 (compare the proof of 2.14) by putting $1, \ldots, 6$ in general position, 7 in the relative interior of $\langle 23456 \rangle$, and 8 in the relative interior of $\langle 1236 \rangle$. The two triangles $\langle 135 \rangle$ and $\langle 246 \rangle$ cover all vertices of the convex hull. By cutting out these triangles we get a tight Klein bottle with two holes by 2.23.

A tight projective plane with one handle and one hole in E^5 can be chosen as a 5-top-set of an embedding of the projective plane with handle into E^6: Consider the triangulation given in Figure 4 and cut out the open star of the vertex 6. The same procedure is possible (even in E^6) for the Klein bottle with one handle if we take an appropriate triangulation containing a K_8 and then cut out the star of an appropriate vertex.

It remains only to cut out additional holes and to attach additional handles. This covers the nonorientable case.

The non-existence results claimed in 2.24 follow by similar arguments: The cases of a disc in E^3, a Möbius band in E^5 and a Klein bottle with one hole in E^5 are excluded by 2.23 because otherwise there would be a tight embedding of S^2 into E^4 or of $\mathbf{R}P^2$ or the Klein bottle into E^6, contradicting 2.15. The Möbius band with one hole in E^5 is excluded because two holes are not sufficient to cover all six vertices of a $K_6 \subset \mathbf{R}P^2$, compare the proof of 2.17 (ii). Surfaces of higher genus with one boundary component cannot be immersed into E^2 because the boundary of the convex hull must be one such component, and there must be another one mapped into the interior.

2.25 Theorem ([Kü2;Thm. C]): *Let M be a compact surface without boundary. There exists a substantial tight polyhedral embedding of M minus an open disc into E^d if and only if*

$$d = 2 \quad \text{if } M \cong S^2,$$

$$d = 3 \text{ or } 4 \quad \text{if } M \text{ is the Klein bottle },$$

$$3 \leq d \leq \frac{1}{2}\left(3 + \sqrt{49 - 24\chi(M)}\right) \quad \text{otherwise.}$$

The upper bound follows from 2.14 in connection with 2.23. The construction for attaining the highest possible codimension in the general case requires the same method as in the proof of 2.14. Moreover, particular attention is required for the position of the boundary with respect to the embedding $K_{d+1} \rightarrow M$. For the details we refer to [Kü2].

2.26 Corollary (Tight triangulations): *Let M be as in 2.25, and let M_1 denote the surface M minus an open disc. Assume that M_1 is triangulated with n vertices. Then the following conditions are equivalent:*

(i) The triangulation is tight.

(ii) The triangulation is 2-neighborly and each vertex is contained in ∂M_1.

(iii) $n = \frac{1}{2}\left(5 + \sqrt{49 - 24\chi(M)}\right)$.

This follows directly from 2.16, 2.23, and 2.25.

For r being sufficiently large one can achieve for M_r the same upper bound as for M (see [Kü2;Thm. D]). However, it seems to be difficult to determine the exact number of holes which is needed to cover all vertices of $K_{d+1} \subset M$. Compare the case $K_6 \subset \mathbf{R}P^2$ where three holes are needed to cover six vertices.

For 0-tight surfaces with boundary the upper bound for the substantial codimension is always the same as for the corresponding surface without boundary (compare 2.14 and [Kü2;Thm. B]). In this case by 2.4 the procedure of cutting out planar convex discs does preserve the 0-tightness.

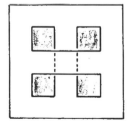 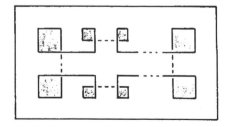

tight orientable surfaces in E^2

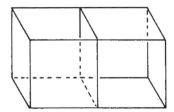 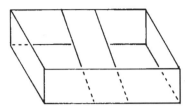

tight torus with one hole   and tight Klein bottle with one hole

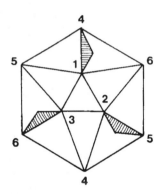 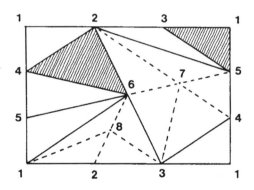

$\mathbf{R}P^2$ with three holes   and Klein bottle with two holes

Figure 5

2F. Surfaces with singularities

In this section we give a few examples of tight polyhedral surfaces with certain types of singularities.

2.27 Definition: An abstract surface with isolated singularities is an ordinary surface (without boundary) after subsequent identification of a finite number of pairs of points. Such an exceptional point is called a proper singularity (or pinch point). A typical neighborhood of such a proper singularity looks like the cone over a disjoint union of circles. In the polyhedral case each such singular point is required to be a vertex. A triangulated surface with isolated singularities is a simplicial complex whose underlying set is a surface with isolated singularities. It is called strongly connected if any two triangles can be joined by a chain of triangles, where each pair of subsequent triangles has an edge in common. For examples see the following Figure 6.

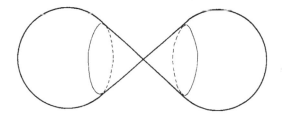

not strongly connected

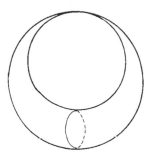

a pinched sphere

Figure 6

The following lemma says that the implication (i) $\Rightarrow$ (ii) in 2.13 remains true for strongly connected surfaces with isolated singularities (compare [Sar2]):

2.28 Lemma: *If there is an embedding $K_n \to M$ then Heawood's inequality $\chi(M) \leq n(7 - n)/6$ holds. In the case of equality this embedding of K_n induces a triangulation of M.*

The proof in 2.13 can be carried over directly. Note, however, that this is not true if M is not strongly connected. In this case it may happen that one of the components of $M \setminus K_n$ is an abstract 2-gon. Therefore the proof in 2.13 would break down. In fact, if M is a union of two 2-spheres where the two north poles and the two south poles are identified, then $\chi(M) = 2$ but M contains a K_5, in disagreement with Heawood's inequality.

Unfortunately, even if equality in Heawood's inequality holds, the triangulation is not a tight triangulation if there are singular vertices. This is easy to see: For a strongly connected surface M with isolated singularities $H_2(M; \mathbf{Z}_2)$ is 1-dimensional. On the other hand for a height function attaining its maximum at a singularity we have $\mu_2 \geq 2$ because this maximum has multiplicity two in the sense of Def. 3.13. This contradicts 2.5 (ii), and it shows that 2.5 is not valid for surfaces with singularities . The same argument as in the proof of 2.23 shows that tightness is equivalent to the equation $\mu_2 = 1$ for every height function with finitely many critical points.

Consequently, in order to construct tight embeddings in this case, one has to choose the positions of the singular vertices more carefully and, in particular, in the interior of the convex hull. We illustrate this phenomenon by the following examples:

2.29 Proposition: *There is a tight polyhedral embedding of a pinched torus into E^4, and there is a tight polyhedral embedding of a pinched surface of genus 2 into E^6.*

PROOF: Let us consider the triangulations in Figure 7. In the first case, the pinched torus, we put the vertices 1, 2, 4 as vertices of a regular triangle in a plane E_1 and 3, 5, 6 as vertices of another regular triangle in a plane E_2. Assume that the origin is an interior point of each of the two triangles in each plane. Then we take the orthogonal sum of E_1 and E_2 intersecting at the origins of each plane. We put the vertex x into the center $E_1 \cap E_2$. The last vertex 0 gets a position near the origin but in general position with respect to all the other vertices. All the edges and triangles can be added without self-intersections. The 0-tightness follows easily from 2.3 and 2.4. For the tightness we have to show $\mu_2(z) = 1$ for any height function in general position. Since the singular vertex is the origin, this just means that the singular vertex is not critical of index 2.

This in turn follows from its special position: no hyperplane can separate x from three vertices forming a cycle in the link of x.

In the case of the pinched surface of genus 2 the triangulation in Fig. 7 comes from a block design, compare [ABS1;Fig.5]. In this case we proceed similarly: Put the vertices 1, 3, 5, 7 as the vertices of a regular tetrahedron in a 3-space E containing the origin in its interior. Furthermore put 2, 4, 6, 8 as the vertices of another regular tetrahedron in another 3-space E'. Then take the orthogonal sum of these 3-spaces, intersecting at the common origin which is the position of the additional singular vertex y. The tightness follows by the same argument as in the case of the pinched torus.

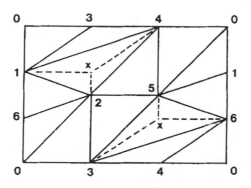

a pinched torus

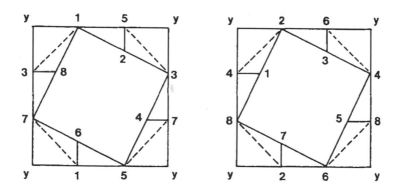

a pinched surface of genus 2

Figure 7

2.30 Corollary: *There is a tight polyhedral immersion of the torus into E^4 with exactly one double point.*

This follows from the construction in 2.29. If we disregard the vertex x as a vertex and introduce the triangles $\langle 124 \rangle$ and $\langle 356 \rangle$ then we obtain exactly the 7-vertex triangulation of the torus. Now the embedding from 2.29 becomes an immersion with exactly one double point ($=$ the origin) as the intersection of these triangles $\langle 124 \rangle$ and $\langle 356 \rangle$, the ones which are invariant under the automorphism $(124)(365)$ of order three. It is clearly tight because the edge graph is a K_7. A smoothing of this immersion has non-vanishing normal Euler class. Therefore a tube around it cannot be tight with respect to a field of characteristic 0, see [Bu-Kü]. It would be interesting to explicitly evaluate the combinatorial formula for the normal Euler class given in [Ba7].

An analogue of 2.14 or 2.15 for pinched surfaces would require a solution of the Heawood problem: find an embedding of K_n into M whenever the Heawood inequality is satisfied. To our knowledge, this is not yet completed. There are, however, many constructions of triangulated pinched surfaces with a complete edge graph K_n. In terms of block designs such a triangulation is nothing but a block design $S_2(2, 3; n)$ without repeated blocks [Wh2]. In this case we have to interpret the abstract triples (blocks) as the triangles of a simplicial complex. The link of each vertex is a union of polygonal circuits. However, one has to decide the question of strong connectedness separately. Constructions can be found in the literature on block designs [BJL]. All such triangulations with $n \leq 10$ are classified, compare [MR], [CCHR]. Polyhedral embeddings into Euclidean 3-space without additional vertices are given in [ABS1] as polyhedra without diagonals.
In principle, the tightness can be dealt with as in 2.29, but the details may be quite complicated. It seems that this work has to be done in the future.

QUESTION: *Assume that M is a given surface with isolated singularities and assume that the equality $\chi(M) = n(7 - n)/6$ is satisfied. In which cases is there a triangulation with n vertices?*
For surfaces without singularities the only exception is the Klein bottle, see 2.13. The Klein bottle with three pinch points is another exception: For $n = 9$ equality $\chi(M) = n(7-n)/6$ is satisfied but there is no such combinatorial structure. This follows from the enumeration of block designs in [MR]. The torus with three pinch points does occur, as already mentioned in [Em;Fig.2], compare also the pinched surface of genus 2 in Figure 7.

Another type of surfaces with singularities is the type of *open-book-surfaces*. In this case one allows singular edges, i.e., edges which are contained in more than two triangles. Along such an edge the surface looks like an open book with more than two pages. The following example shows that tightness is possible for such open-book-surfaces, moreover it shows that the field of coefficients is essential.

2.31 Theorem ([Kü6]): *There exists a 2-dimensional simplicial complex K such that the following hold:*

1. *K is a 2-manifold except along a cycle of three edges (an open-book-surface).*
2. *The first integral homology $H_1(K)$ has 3-torsion.* 3. *There is a polyhedral immersion $f: K \to E^4$ which is $\mathbf{Z}_3$-tight but not $\mathbf{Z}_2$-tight.*

PROOF: We define K as the simplicial complex built up by 12 vertices $1, 2, 3, 4, 5, 6, 7, 8, 9, x, y, z$ and 45 triangles as follows:

123	456	789	159	267	348	147	258	369
xy1	yz4	zx7				128	452	785
xy2	yz5	zx8				239	563	896
xy3	yz6	zx9				317	641	974
x57	y81	z24				x15	y48	z72
x68	y92	z35				x26	y59	z83
x49	y73	z16				x34	y67	z91

It is invariant under the $(\mathbf{Z}_3 \times \mathbf{Z}_3)$-action generated by the permutations

$$\alpha = (123)(456)(789)$$

and

$$\beta = (147)(258)(369)(xyz).$$

K is a 2-manifold except along the edges xy, yz, zx where three triangles meet. K is covered by the orientable surface with boundary shown in Figure 8. The three boundary components $x_1 y_1 z_1, x_2 y_2 z_2, x_3 y_3 z_3$ have to be identified. The integral homology of K is the following:

$$H_0(K) \cong \mathbf{Z}, \quad H_1(K) \cong \mathbf{Z}^{10} \oplus \mathbf{Z}_3, \quad H_2(K) = 0.$$

With respect to the field $\mathbf{Z}_3$ the Betti numbers are $b_0(K) = b_2(K) = 1, b_1(K) = 11$. Therefore, any function in general position has at least 13 critical points, counted with multiplicity (Def. 3.13). As in the beginning of this Section 2F, each of the exceptional points x, y, z contributes $\mu_2 = 2$ if it is a maximum of a height function. Therefore this triangulation is not a tight triangulation, and these three vertices should lie in the interior of the convex hull of the others, otherwise tightness has no chance, as explained above. We define a simplexwise linear mapping of K into 4-space by the following positions of the 12 vertices:

$$1 \mapsto (2,0,2,0) \qquad 4 \mapsto (-1,\sqrt{3},2,0) \qquad 7 \mapsto (-1,-\sqrt{3},2,0)$$
$$2 \mapsto (2,0,-1,\sqrt{3}) \qquad 5 \mapsto (-1,\sqrt{3},-1,\sqrt{3}) \qquad 8 \mapsto (-1,-\sqrt{3},-1,\sqrt{3})$$
$$3 \mapsto (2,0,-1,-\sqrt{3}) \qquad 6 \mapsto (-1,\sqrt{3},-1,-\sqrt{3}) \qquad 9 \mapsto (-1,-\sqrt{3},-1,-\sqrt{3})$$
$$x \mapsto (\varepsilon\sqrt{3},\varepsilon,0,0) \qquad y \mapsto (-\varepsilon\sqrt{3},\varepsilon,0,0) \qquad z \mapsto (0,-2\varepsilon,0,0)$$

where ε denotes a sufficiently small positive number. This mapping is not yet an immersion. We have to perturb it a little bit until all vertices lie in general position. This immersion is tight because $\mu_0 = \mu_2 = 1$ for every height function in general position. The special position of the singular vertices x, y, z implies that they are not critical of index 2. In fact, no hyperplane can separate any of the singular vertices from any set of vertices forming a cycle in the link.

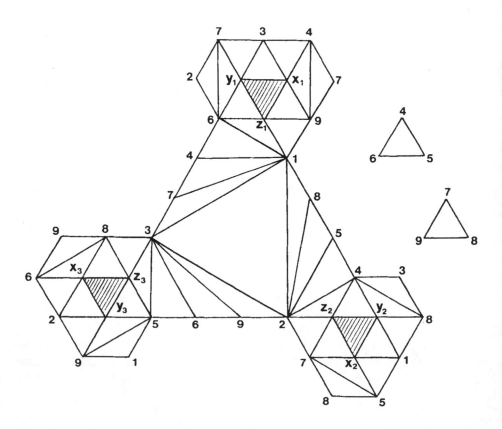

Figure 8

3. Tightness and k-tightness

The 0-tightness of a compact polyhedron in E^d can be expressed by saying that no height function has a disconnected set of local minima or maxima. In particular, for a height function with isolated extrema there is exactly one minimum and one maximum in this case. In terms of the 0th homology this means that a 0-dimensional cycle lying in a halfspace already bounds in this halfspace if it bounds in M. In order to deal with 1-tightness, 2-tightness etc. and tightness in general we have to talk about the same matter for 1-cycles, 2-cycles etc.

It is also very convenient to define an appropriate critical point theory for a certain class of 'nondegenerate' real functions which corresponds to the class of Morse functions in differential topology. An outline of such a critical point theory together with several consequences is contained in Section 3B. As an application, we describe in Section 3C a general method of constructing tight polyhedra of arbitrary dimension in arbitrarily high dimensional space.

3A. Tightness of higher dimensional polyhedra

3.1 Definition: A compact polyhedron $M \subset E^d$ is called k-tight with respect to a field F if for every open or closed half space $h \subset E^d$ the induced homomorphism

$$H_i(h \cap M; F) \to H_i(M; F)$$

is injective for $i = 0, 1, \ldots, k$. It is called tight with respect to F (or F-tight) if this holds for all i.

H_i denote the ith singular homology group which in our case is an F-vectorspace. Since $h \cap M$ has the homotopy type of a finite polyhedron, the simplicial homology theory is also appropriate for all considerations in this volume. The injectivity of the induced homomorphism for a fixed half space h and a fixed dimension i just means that any homology i-cycle of M which lies in this half space does not bound in M whenever it does not bound in this half space.

If M is a connected polyhedron then the 0-tightness is equivalent to Banchoff's Two-piece-property (TPP) saying that $h \cap M$ is always connected. Similarly, if M is connected and simply connected then the 1-tightness is equivalent to the property that $h \cap M$ is always connected and simply connected. In general: if M is k-connected then the k-tightness is equivalent to the property that $h \cap M$ is always k-connected.

Examples of tight polyhedra are convex polytopes, their boundaries, and their k-dimensional skeletons. Tight polyhedral surfaces are discussed in Chapter 2. Five of the six squares of a cube form a 2-dimensional polyhedron which is 0-tight but not 1-tight. Similarly, seven of the eight 3-faces of a 4-cube form a polyhedron which is 1-tight but not 2-tight.

3.2 Definition (Canonical embedding, tight triangulation): Let K be a simplicial complex with n vertices and let $\triangle^{n-1}$ denote the simplex spanned by n points in E^{n-1} in general position.

(i) $K \subset \triangle^{n-1} \subset E^{n-1}$ is called the canonical embedding of K. Its image is unique up to affine transformations of E^{n-1}.

(ii) K is called a tight (or k-tight) simplicial complex if the canonical embedding is tight (or k-tight, respectively). In this case we say that K is a tight (or k-tight) triangulation if K is a triangulated manifold with or without boundary.

3.3 Definition: Let M be a subcomplex of the boundary complex of a convex polytope P. M is called k-Hamiltonian in P if M contains $Sk_k(P)$. This generalizes the notion of a 1-Hamiltonian subcomplex in 2.11. The classical concept of a Hamiltonian circuit is nothing but a 0-Hamiltonian subcomplex homeomorphic to the circle.

The following proposition gives a higher dimensional generalization of 2.10.

3.4 Proposition: *Let $M \subset E^d$ be a $(k-1)$-tightly embedded $(k-1)$-connected polyhedron. Then M contains the k-skeleton of its convex hull $\mathcal{H}$. The same holds under the assumption that $H_i(M) = 0$ for $i = 1, \ldots, k-1$.*

PROOF: For $k = 1$ the assertion holds by 2.4. Now assume that M contains $Sk_i(\mathcal{H})$ for some i, $1 \leq i < k$, and let A be an $(i+1)$-face of $\mathcal{H}$. There is a half space h such that $A = h \cap \mathcal{H}$. By assumption $h \cap M$ contains the i-dimensional boundary of A which is homologous to zero in M if we regard it as an i-cycle. The i-tightness implies $h \cap M = A$. Therefore M contains $Sk_{i+1}(\mathcal{H})$. Inductively it follows that M contains $Sk_k(\mathcal{H})$.

3.5 Lemma (B. Grünbaum): *The k-skeleton of any convex d-polytope contains the k-skeleton of the d-simplex as a subset (not necessarily as a subcomplex).*

PROOF: This is a straightforward generalization of 2.8. The inductive proof of 2.8 for the case $k = 1$ can be carried over directly to the case of arbitrary k, for the details see [Grü;11.1].

3.6 Corollary: *Let $\Sigma^n \to E^d$ be a tight and substantial polyhedral embedding of an n-dimensional homology sphere Σ^n. Then $d = n + 1$ and the image is the boundary of a convex d-polytope.*

PROOF: $H_1(\Sigma^n) = \ldots = H_{n-1}(\Sigma^n) = 0$ implies by 3.4 that the image contains $Sk_n(\mathcal{H})$. This in turn contains $Sk_n(\triangle^d)$ by 3.5. On the other hand for $d > n+1$ $Sk_n(\triangle^d)$ contains $Sk_n(\triangle^{n+1}) \cong S^n$ as a proper subcomplex. This is impossible since no compact and connected n-manifold contains S^n as a proper submanifold. Therefore $d = n + 1$ and Σ^n must coincide with $Sk_n(\mathcal{H}) = \partial \mathcal{H}$.

3.7 Corollary: *Let M be a tight triangulation of the sphere S^{n-1} (or of any homology $(n-1)$-sphere). Then M is combinatorially equivalent to the boundary complex of the n-dimensional simplex $\triangle^n$ with $n + 1$ vertices.*

This follows directly from 3.6: Let $d + 1$ be the number of vertices of the tight triangulation and consider the canonical embedding into E^d.

3.8 Corollary: *Let M be a polyhedral subcomplex of the boundary complex of a convex d-polytope P such that $P = \mathcal{H}(M)$. Then (i) and (ii) are equivalent:*

(i) M is $(k-1)$-connected and the embedding $M \subset P \subset E^d$ is $(k-1)$-tight.

(ii) M is k-Hamiltonian in P.

PROOF: (i) $\Rightarrow$ (ii) holds by 3.4. Now assume (ii). Then M contains the $(k-1)$-connected subset $Sk_k(P)$. On the other hand every continuous map $f: S^\ell \to M$ is homotopic to a map $\bar{f}: S^\ell \to Sk_k(M) = Sk_k(P)$ for $\ell \leq k - 1$. Therefore M is $(k-1)$-connected. The same argument holds for $h \cap M$ where h is an arbitrary half space because $h \cap Sk_k(P)$ is also $(k-1)$-connected. This proves the $(k-1)$-tightness.

3.9 Definition (Neighborly and cyclic polytopes): A simplicial d-polytope (or any simplicial complex) with n vertices is called k-neighborly if any k-tuple of vertices spans a $(k-1)$-dimensional simplex of the polytope (or the complex, respectively). An equivalent formulation is that the number of $(k-1)$-dimensional simplices is $\binom{n}{k}$. A simplicial d-polytope is called neighborly if it is k-neighborly for any $k \leq d/2$, see [Grü;7.2].

A particular example is the cyclic d-polytope $C(n,d)$ defined as the convex hull of n subsequent points on the moment curve $t \mapsto (t, t^2, t^3, \ldots, t^d) \in E^d$. $C(n,d)$ is a simplicial d-polytope which is $d/2$-neighborly if d is even and $(d-1)/2$-neighborly if d is odd. It is therefore neighborly. The face structure of $C(n,d)$ is determined by Gale's evenness condition, see [Grü;3.24]. In particular, this structure is independent of the choice of the points on the moment curve. If $d = 2m$ is even, then an alternative description of the cyclic polytope is given by the convex hull of n subsequent points on the trigonometric moment curve $t \mapsto (\cos t, \sin t, \cos 2t, \sin t, \cos 3t, \sin 3t, \ldots, \cos mt, \sin mt) \in E^d$. In particular, $C(7,4)$ is the convex hull of 7 equidistant points $t_k = 2\pi k/7, k = 0, 1, 2, \ldots, 6$ on the curve $(\cos t, \sin t, \cos 2t, \sin 2t)$. The 2-skeleton of $C(7,4)$ contains the 7-vertex triangulation of the torus. This leads to a tight polyhedral embedding of the torus into E^4 where the automorphism of order 7 is realized by a Euclidean rotation, compare 2.6 and 2.19.

3.10 Corollary: *For a simplicial complex K the following conditions are equivalent:*

(i) K is $(k-1)$-connected and $(k-1)$-tight.

(ii) K is $(k+1)$-neighborly.

This is a reformulation of 3.8 for the case of the canonical embedding of K.

3.11 Corollary: *Let d, k, n be given natural numbers satisfying $n \geq d+1$ and $k \leq (d-2)/2$. Then there is a $(k-1)$-tight triangulation of S^{d-1} with n vertices.*

PROOF: According to 3.10 we have to look for $(k+1)$-neighborly triangulations of S^{d-1}. The boundary complex of any neighborly d-polytope is an example, in particular the cyclic d-polytopes $C(n,d)$ provide perfect examples, see 3.9.

0-tight polyhedral embeddings of spheres with codimension greater than 1 are originally due to T. Banchoff [Ba4]. The results of 3.11 and 3.12 are in sharp contrast to the smooth case: If $f: M^n \to E^d$ is a smooth 0-tight and substantial immersion of a compact manifold then $d \leq n(n+3)/2$ [Kui4].

3.12 Theorem (D. Walkup, K.S. Sarkaria, [Wa], [Sar1]): *For any given 3-manifold there is a 2-neighborly (hence 0-tight) triangulation if the number of vertices is chosen sufficiently large.*

In general, no precise information is available about the number of vertices which is necessary. Compare, however, the remark in 5.3.

QUESTION: *Does any given $(k-1)$-connected $(2k+1)$-manifold admit a $(k+1)$-neighborly (hence $(k-1)$-tight) triangulation?*

3B. Polyhedral critical point theory

A basic result in classical Morse theory states that the homotopy type (or homology type) of the sublevelset of a nondegenerate function can change only at levels containing critical points, i.e., points where the differential of the function vanishes. For nonsmooth functions we can use this as a definition of critical points as follows (compare [Kui5], [Ba2], [Ba6]):

3.13 Definition (Critical points): Let F be a fixed field, and let $K \subset E^n$ be a compact polyhedron. For a unit vector $z \in \mathbf{R}^n$ let $z^* \colon K \to \mathbf{R}$ denote the height function

$$z^*(x) := \langle x, z \rangle.$$

If $z^*(p) \neq z^*(q)$ for any pair p, q of vertices then we say that it is a height function in general position. A point $x \in K$ is called critical for z^* if

$$H_*(K_x, K_x \setminus x; F) \neq 0$$

where $K_x := \{ y \in K \mid z^*(y) \leq z^*(x) \}$ and $K_x \setminus x := K_x \setminus \{x\}$. In particular, if z^* is in general position then no point can be critical except for the vertices. Thus the number is finite by assumption. In more detail

$$(\dim_F H_0(K_v, K_v \setminus v), \ldots, \dim_F H_n(K_v, K_v \setminus v))$$

is called the multiplicity vector of a vertex v, and v is said to be critical of index i and multiplicity m if $\dim_F H_i(K_v, K_v \setminus v) = m > 0$. $\sum_i \dim_F H_i(K_v, K_v \setminus v)$ is called the total multiplicity of v, and

$$\mu_i(z; F) \quad := \quad \sum_v \dim_F H_i(K_v, K_v \setminus v; F)$$

$$\mu(z; F) \quad := \quad \sum_i \mu_i(z; F)$$

are called the number of critical points of z^* of index i and the number of critical points of z^*, respectively, each counted with multiplicity. In contrast with the ordinary Morse theory for smooth functions, $\mu_i(z;F)$ does depend on the choice of F; for examples see the remark after 3.23, and 3.25. The case of higher multiplicity $m > 1$ occurs already for height functions on surfaces in E^3, e.g., at generic monkey saddles, see [Ba-T].

We recall the following duality theorem in homology and cohomology with coefficients in a field F, for the details and proofs see [Do;VIII]:

3.14 Proposition (Duality for manifolds:) *Let M denote a compact d-manifold, and let K denote a compact polyhedron. Then the following hold:*

(i) algebraic duality:

$$H^i(K;F) \cong (H_i(K;F))^* \cong H_i(K;F)$$

(ii) Alexander duality for $K \subset S^d$:

$$\tilde{H}_i(K;F) \cong \tilde{H}_{d-i-1}(S^d \setminus K;F)$$

(iii) Poincaré-Lefschetz duality for $K \subset M$:

$$H_i(K;\mathbf{Z}_2) \cong H_{d-i}(M, M \setminus K; \mathbf{Z}_2)$$

(iv) Poincaré duality:

$$H_i(M;\mathbf{Z}_2) \cong H_{d-i}(M;\mathbf{Z}_2)$$

(v) Lefschetz duality for a compact d-manifold M with boundary ∂M:

$$H_i(M;\mathbf{Z}_2) \cong H_{d-i}(M, \partial M; \mathbf{Z}_2).$$

(iii), (iv) and (v) hold for an arbitrary field F provided that M is orientable.

3.15 Proposition (Morse Relations, [Kui5]): *Let $K \subset E^n$ be a compact polyhedron and z^* be a height function in general position. We denote the Betti numbers by $b_i(X;F) := \dim_F H_i(X;F)$. Then the following hold:*

(i) $\mu_i(z;F) \geq b_i(K;F)$ *for each $i = 0, 1, \ldots$*

(ii) $\mu(z;F) \geq \sum_i b_i(K;F)$

(iii) $\sum_{i \leq k} (-1)^{k-i} \mu_i(z;F) \geq \sum_{i \leq k} (-1)^{k-i} b_i(K;F)$ *for each $k = 0, 1, \ldots$*

(iv) $\sum_i (-1)^i \mu_i(z;F) = \chi(M) = \sum_i (-1)^i b_i(K;F)$

(v) $\mu_i(z; F) = b_i(K; F)$ *if and only if for every half space* $h \subset E^n$, *whose boundary is orthogonal to* z, *the induced homomorphism* $H_i(h \cap K; F) \to H_i(K; F)$ *is injective.*

(vi) K *is tight with respect to* F *(or k-tight) if and only if* $\mu_i(z; F) = b_i(K; F)$ *for every height function in general position and for each* i *(or for* $i = 0, 1, \ldots, k$, *respectively).*

(vii) *If* K *is a d-manifold without boundary then* $\mu_i(z; F) = \mu_{d-i}(-z; F)$ *for* $i = 0, 1, \ldots, d$.

(i) - (vii) hold also for abstract simplexwise linear functions on a simplicial complex attaining different values at distinct vertices, compare [Kü5]. Functions satisfying equality in (ii) are often called <u>tight</u> or <u>perfect functions</u>.

PROOF: (i) Let $v_1, \ldots, v_m$ be the critical points of z^* such that

$$z^*(v_1) < z^*(v_2) < \ldots < z^*(v_m),$$

and let $K_j := K_{v_j}$.

From the long exact homology sequence

$$\ldots \longrightarrow H_i(K_j) \longrightarrow H_i(K_{j+1}) \longrightarrow H_i(K_{j+1}, K_j) \longrightarrow H_{i-1}(K_j) \longrightarrow \ldots$$

we get $b_i(K_{j+1}) \le b_i(K_j) + \dim H_i(K_{j+1}, K_{j+1} \setminus v_{j+1})$.

Inductively this implies

$$b_i(K) = b_i(K_m) \le b_i(K_{m-1}) + \dim H_i(K_m, K_m \setminus v_m)$$

$$\le b_i(K_{m-2}) + \dim H_i(K_{m-1}, K_{m-1} \setminus v_{m-1}) + \dim H_i(K_m, K_m \setminus v_m)$$

$$\le \sum_j \dim H_i(K_j, K_j \setminus v_j) = \mu_i(z).$$

(ii) is an obvious consequence of (i), (iii) follows similarly from the long exact sequence above by taking alternating sums, as in classical Morse theory. (iv) is a direct consequence of (iii).

(v) We assume $\mu_i(z) = b_i(K)$ for a fixed height function z^* and a fixed index i. From the proof of (i) we see that this equality implies

$$b_i(K_{j+1}) = b_i(K_j) + \dim H_i(K_{j+1}, K_{j+1} \setminus v_{j+1})$$

and that the induced homomorphism $H_i(K_j) \to H_i(K_{j+1})$ is injective for every j. Therefore $H_i(K_j) \to H_i(K)$ is injective for every j. On the other hand, for any half space h whose boundary is orthogonal to z, it is true that $h \cap K$ retracts onto some K_j by a deformation retract. Therefore $H_i(h \cap K) \to H_i(K)$ is injective for any such h.

(vi) follows from (v).

(vii) By assumption for each vertex the vertex star, denoted by star(v), is a d-ball, and the vertex link, denoted by link(v), is its boundary $(d-1)$-sphere. Let $K_v(z) := \{x \in M \mid z^*(x) \le z^*(v)\}$. A vertex v is critical of index i for the height function z^* if

$$H_i(K_v(z), K_v(z) \setminus v) \cong H_i(K_v(z) \cap \text{star}(v), K_v(z) \cap \text{link}(v))$$

$$\cong \tilde{H}_{i-1}(K_v(z) \cap \text{link}(v)) \ne 0.$$

For the opposite height function $-z^*$ we get

$$H_{d-i}(K_v(-z), K_v(-z) \setminus v) \cong H_{d-i}(K_v(-z) \cap \text{star}(v), K_v(-z) \cap \text{link}(v))$$

$$\cong \tilde{H}_{d-i-1}(K_v(-z) \cap \text{link}(v)) \cong \tilde{H}_{d-i-1}(\text{link}(v) \setminus K_v(z)).$$

By the Alexander duality 3.14 (ii) for link(v) $\cong S^{d-1}$ we get

$$\tilde{H}_{i-1}(K_v(z) \cap \text{link}(v)) \cong \tilde{H}_{d-i-1}(K_v(-z) \cap \text{link}(v))$$

and therefore $\mu_i(z) = \mu_{d-i}(-z)$.

3.16 Proposition (Tightness of products):

(i) *If $K \subset E^n$ is tight and $E^n \to E^m$ is a linear mapping then $K \to E^n \to E^m$ is also tight.*

(ii) *If $K \subset E^n$ and $L \subset E^m$ are tight then $K \times L \subset E^n \times E^m \cong E^{n+m}$ is also tight.*

(i) is an obvious consequence of the definition, (ii) follows from the Künneth formula for the local homology around a vertex in the cartesian product. In more generality, using Čech homology T.Ozawa [Oz] showed that products of tight subsets are again tight.

3.17 Proposition: *Let K be a (finite) simplicial complex with n vertices. Then the following conditions are equivalent:*

(i) *K is tight (Def. 3.2),*

(ii) *every simplexwise linear embedding of K into any Euclidean space is tight,*

(iii) *for any subset V of the set of vertices the induced homomorphism*

$$H_*(\langle V \rangle \cap K) \to H_*(K)$$

is injective where $\langle V \rangle$ denotes the simplex spanned by V,

(iv) *every simplexwise linear function $f: K \to \mathbf{R}$ which satisfies $f(v) \neq f(v')$ for any pair $v \neq v'$ of vertices is a tight function, i.e., it has the minimum number of critical points (counted with multiplicity) which equals $\Sigma_i b_i(K; F)$.*

PROOF: (i) $\Rightarrow$ (ii) is a direct consequence of 3.16 (i) because any simplexwise linear embedding of K is just the canonical embedding followed by a linear projection.
(ii) $\Rightarrow$ (i) is trivial.
(i) $\Leftrightarrow$ (iv) follows from 3.15 (v) since every abstract simplexwise linear function on K can be interpreted as a height function of the canonical embedding of K. Finally, (i) $\Leftrightarrow$ (iii) follows from the fact that for every half space $h \subset E^{n-1}$ the intersection $h \cap K$ has the same homotopy type as $\langle V \rangle \cap K$ where V is the set of vertices in $h \cap K$, and conversely, that $\langle V \rangle \cap K$ is a top-set of the canonical embedding.

3.18 Proposition (Duality for tightness): *For any closed $(2k-1)$-manifold or $2k$-manifold M which is embedded into Euclidean space (polyhedrally or smoothly) the following conditions are equivalent:*

(i) M is tight with respect to $\mathbf{Z}_2$.

(ii) M is $(k-1)$-tight with respect to $\mathbf{Z}_2$.

Consequently, a triangulation of M is tight if and only if it is $(k-1)$-tight (for the field $F = \mathbf{Z}_2$).

PROOF: (i) $\Rightarrow$ (ii) holds by definition.
Now assume (ii) and consider a height function z^* which is in general position (if M is polyhedral) or which is a Morse function (if M is smooth). Let n denote the dimension of M. By assumption $\mu_i(z) = b_i(M)$ for $i = 0, \ldots, k-1$. By the duality 3.15 (vi) and by Poincaré duality 3.14 (iv) we have $\mu_{n-i}(z) = \mu_i(-z) = b_i(M) = b_{n-i}(M)$. Finally the Euler-Poincaré relation 3.15 (iii) implies $\mu_k(z) = b_k(M)$ if $n = 2k$.
For other fields F different from $\mathbf{Z}_2$ this argument depends on the validity of the Poincaré duality. Note that 3.18 is not true for manifolds with boundary. The cone over any tightly embedded $(n-1)$-sphere provides an example of an $(n-2)$-tight n-manifold-with-boundary which is not tight. An embedding of an n-manifold with boundary is tight if and only if it is $(n-2)$-tight and in addition $\mu_n(z) = 0$ for every height function in general position.

3C. Tight subcomplexes of the cube

In this section we describe a general method how to construct tight polyhedra (manifolds or not) as subcomplexes of higher dimensional cubes, generalizing the construction of tight surfaces in 2.12. It turns out that any simplicial complex K with n vertices induces in a canonical way a tight subcomplex 2^K of the n-dimensional cube C^n.

3.19 Definition (L. Danzer's construction, [Dz], [MM-S]): Let K be a simplicial complex with d vertices $1, 2, \ldots, d$. Each k-simplex of K can be identified with a subset $\triangle = \{i_0, \ldots, i_k\}$ of $\{1, \ldots, d\}$. Let us define

$$A_j(\triangle) \quad := \begin{cases} [0,1] & \text{if } j \in \{i_0, \ldots, i_k\} \\ \{0,1\} & \text{otherwise} \end{cases}$$

$$F(\triangle) \quad := A_1(\triangle) \times \ldots \times A_d(\triangle)$$

$$2^K \quad := \bigcup_{\triangle \in K} F(\triangle)$$

By definition we may regard each $F(\triangle)$ and therefore the entire 2^K as a subcomplex of the d-dimensional cube $C^d := [0,1]^d$ as well as a subset $2^K \subset C^d \subset E^d$ of the ambient Euclidean space.

EXAMPLES: Several simple cases of K and 2^K are listed in the following table. The notation $\{3^d\}$ refers to the Schläfli symbol for regular polytopes [Cox-M].

K	2^K
d isolated vertices	$Sk_1(C^d)$
$\triangle^{d-1}$	C^d
$\partial\triangle^{d-1} = \{3^{d-2}\}$	$\partial C^d = \{4, 3^{d-2}\}$
$Sk_{n-1}(\triangle^{d-1})$	$Sk_n(C^d)$
regular d-gon $\{d\}$	surface of genus g_d (see 2.12)
cross-polytope $\{3^{d-2}, 4\}$	$\partial C^2 \times \ldots \times \partial C^2 \subset C^{2d}$ (d copies)

3.20 Lemma:

(i) 2^K *is an* n-*dimensional manifold without (or with) boundary if and only if* K *is an* $(n-1)$-*dimensional triangulated sphere (or ball, respectively). Moreover,* 2^K *carries an induced PL structure if and only if* K *is a combinatorial sphere (or ball, repectively).*

(ii) 2^K *is* k-*Hamiltonian in* C^d *if and only if* K *is* k-*neighborly with* d *vertices.*

(iii) $2^{K*L} = 2^K \times 2^L$ *where* $*$ *is the join operator for simplicial complexes.*

PROOF: (i) follows from the fact that the link of each vertex of 2^K is combinatorially isomorphic to K.
(ii) and (iii) are obvious by construction.

3.21 Corollary: *Let* K *be a* k-*neighborly combinatorial sphere with* d *vertices. Then* $2^K \subset E^d$ *is a* $(k-1)$-*connected and* $(k-1)$-*tight submanifold.*

The proof follows directly from 3.8 and 3.20.

3.22 Lemma (U. Brehm, [Br4]): *Let F be a field. Then the homology of 2^K can be expressed in terms of the homology of subsets $L \subset K$ as follows:*

$$\tilde{H}_i(2^K; F) \cong \bigoplus_L \tilde{H}_{i-1}(L; F)$$

where $\tilde{H}_i$ denotes the ith reduced homology and where the sum ranges over all non-empty full subcomplexes L of K. A subcomplex is called full if it is the span of a given subset of vertices (following [RS;Ch.3]).

The proof of 3.22 is a longer computation in homological algebra.

3.23 Theorem (Tightness of 2^K): *Let K be any simplicial complex with d vertices. Then $2^K \subset C^d \subset E^d$ is tight with respect to any field F (in fact, 2^K is a tight subcomplex of C^d in the sense of 1.4.*

PROOF: Let $M := 2^K$ and let $z^*: M \to \mathbf{R}$ denote any height function in general position. By 3.15 the tightness of $M \subset E^d$ is equivalent to the equality

$$\sum_{v,i} \dim_F H_i(M_v, M_v \setminus v) = \sum_j \dim_F H_j(M)$$

where the sum ranges over all vertices v of M.
On the other hand by the excision property

$$H_i(M_v, M_v \setminus v) \cong \tilde{H}_{i-1}(L_v)$$

where $L_v := M_v \cap \mathrm{link}\,(v)$ is a full subcomplex of the link of v in M, the latter being combinatorially isomorphic to K. By the Boolean algebra structure on the set of vertices and edges of the cube, L_v ranges over all 2^d such subcomplexes of K when v ranges over all 2^d vertices of M. Therefore 3.22 implies

$$\sum_v \dim_F \tilde{H}_{i-1}(L_v) = \dim_F H_i(M) \text{ for } i \geq 1$$

and finally

$$\sum_v \dim_F H_0(M_v, M_v \setminus v) = \dim_F H_0(v_0, \emptyset) = 1 = \dim_F H_0(M)$$

for the absolute minimum v_0 of z^*.

REMARK: Note that in general the sum of the Betti numbers does depend on the choice of F. If there is p-torsion and q-torsion simultaneously in 2^K, then for $F = \mathbf{Z}_p$ the q-torsion will be ignored by the homology of 2^K but also by the height functions, as far as the number of critical points is concerned. This is different in the case of smooth immersions because the number of critical points of a classical Morse function does not depend on the choice of the field of coefficients.

3.24 Corollary: *Let K be a combinatorial $(n-1)$-sphere with d vertices. Then $2^K \subset C^d \subset E^d$ is a tightly embedded polyhedral n-manifold.*

This follows directly from 3.23 and 3.20 (i).

If K and K' are combinatorially distinct then 2^K and $2^{K'}$ are geometrically distinct. For the huge number of combinatorially distinct triangulations of the sphere see [Kal2]. In some sense 3.23 says that there are more tight polyhedra than simplicial complexes and 3.24 says that there are more tight polyhedral submanifolds than combinatorial spheres. For the topological type of 2^K for special kinds of triangulated spheres K (stacked, neighborly, cyclic) see [Kü-Sch].

3.25 Corollary (Tightness and torsion): *Let K be a combinatorial $(n-1)$-sphere with d vertices containing a full subcomplex $L \subset K$ with non-vanishing p-torsion in the homology for some prime p. Then $2^K \subset E^d$ is a tight n-dimensional submanifold with non-vanishing p-torsion in the homology.*

This follows directly from 3.22 and 3.24.

For smooth taut immersions torsion turns out to be a kind of obstruction: a theorem of G. Thorbergsson [Th2] says that the first non-vanishing homology group has no p-torsion for $p \neq 2$. No example seems to be known for a tight smooth immersion of a manifold with p-torsion for $p \neq 2$.

3.26 Definition (polyhedral ϵ-tube): For a finite polyhedron $K \subset E^d$ and for given $\epsilon > 0$ let

$$K_\epsilon := \{x \in E^d \mid \text{dist}\,(x, K) \leq \epsilon\}$$

be the ϵ-tube around K where the distance function

$$\text{dist}\,(x, y) := \|x - y\|_{\max}$$

is taken with respect to the maximum norm. For any sufficiently small ϵ it is clear that K_ϵ is a polyhedral d-manifold with boundary.

3.27 Theorem (Tightness of polyhedral tubes, [Kü6]): *Let K be a subcomplex of the d-cube and let $0 < \epsilon < \frac{1}{2}$ be a given number. Assume that K is tight with respect to F. Then the following hold:*

(i) *K_ϵ is tight with respect to F.*

(ii) *$\partial(K_\epsilon)$ is tight with respect to F whenever $\sum_i b_i(\partial(K_\epsilon); F) = 2 \cdot \sum_i b_i(K; F)$.*

(iii) *If K is n-dimensional and if $2n + 1 < d$, then $\partial(K_\epsilon)$ is tight with respect to F.*

PROOF: (i) Since K and K_ϵ have the same homotopy type it is sufficient to show that the number of critical points of a height function in general position does not change when passing from K to K_ϵ. By construction each vertex v of K leads to at most 2^d vertices of K_ϵ, namely the vertices of the cube with edge length 2ϵ centered at v (this coincides with the ϵ-ball around v with respect to dist). It is easily seen that none of those vertices is critical except the one with the lowest level and that the lowest one contributes exactly the same total multiplicity as v did before (just as for the ordinary Euclidean ϵ-tube in the smooth case).

(ii) Since K_ϵ is a d-manifold in E^d there are not critical points in the interior. Furthermore the number of critical points on $\partial(K_\epsilon)$ is exactly twice the number of critical points on K_ϵ (or K), see [Kü1;4.2].

(iii) follows from (ii) by verifying the equation

$$\sum_i b_i(\partial(K_\epsilon); F) = 2 \cdot \sum_i b_i(K; F).$$

∂K_ϵ is an orientable hypersurface of E^d, hence it satisfies the Poincaré duality

$$b_i(\partial K_\epsilon) = b_{d-1-i}(\partial K_\epsilon)$$

for any F. Furthermore by the Lefschetz duality 3.14 (v)

$$H_{d-i}(K_\epsilon, \partial(K_\epsilon)) \cong H_i(K)$$

the long exact sequence for the pair $(K_\epsilon, \partial(K_\epsilon))$ can be rewritten as

$$\ldots \to H_i(\partial K_\epsilon) \to H_i(K_\epsilon) \to H_{d-i}(K_\epsilon) \to H_{i-1}(\partial K_\epsilon) \to \ldots$$

where by assumption

$$H_i(K_\epsilon) = H_i(K) = 0 \quad \text{for } i > n.$$

This means that in this sequence for each i either $H_i(K_\epsilon)$ or $H_{d-i}(K_\epsilon)$ vanishes. This implies the assertion independently of the field F.

REMARKS: 1. The assumption in (ii) is not satisfied if K is an immersed nonorientable hypersurface. For a smooth $\mathbf{Z}_2$-tight embedding the tube is always $\mathbf{Z}_2$-tight. Its boundary is $\mathbf{Z}_2$-tight if and only if the mod 2 Euler class of the normal bundle vanishes, see [Bu-Kü] including an analogue of 3.27 in the smooth case. For taut immersions, the boundary of the tube is always taut, see [Pi2].
Note that the polyhedral tube is adapted particularly to subcomplexes of the cube. In general, the same construction for arbitrary polyhedra would not lead to tightness, and 3.27 would not hold.

2. In some sense 3.27 together with 3.23 says that there are more tight hypersurfaces than simplicial complexes because for every simplicial complex K with sufficiently many vertices (compared with the dimension) a tight hypersurface is given by $\partial(2^K)_\epsilon$. A particular example illustrating 3.27 is given by the k-skeleton of C^d whose boundary of the polyhedral ϵ-tube is PL homeomorphic to a connected sum of copies of $S^k \times S^{d-k-1}$. For an explicit example with 3-torsion see 2.31 or [Kü6]. More sophisticated topological properties of K will, in general, lead to special or even 'pathological' properties of 2^K. The following proposition gives rather peculiar examples.

3.28 Proposition:

(i) There is a tight polyhedral homology 4-manifold in E^d (for some $d \geq 17$) which is not a manifold.

(ii) There is a tight polyhedral complex in E^d (for d sufficiently large) which is homeomorphic to a simply connected compact 6-manifold but not PL with respect to the induced polyhedral decomposition.

PROOF: (i) Let K be a triangulation of a homology 3-sphere, e.g., the Poincaré dodecahedron space. A triangulation with 17 and a more symmetric one with 18 vertices were constructed by U. Brehm (personal communication). It follows that 2^K is a homology 4-manifold which is tight by 3.23. It is not a manifold because the links of the vertices are not simply connected.

(ii) Here we start with the Edwards double suspension $S^2(\Sigma^3)$ of a certain homology 3-sphere Σ^3 which is homeomorphic to S^5 but not PL, see [Ed]. Let K' be a 2-neighborly triangulation of Σ^3 with $d-4$ vertices. The double suspension ($=$ two-fold double cone) introduces four extra vertices x_1, x_2, y_1, y_2 where the edges $x_1 x_2$ and $y_1 y_2$ are missing. By a slight modification we can introduce these edges without changing the PL type. This yields a 2-neighborly triangulation K of the Edwards sphere with d vertices which is still not PL.
Consequently 2^K is 2-Hamiltonian in C^d and hence simply connected by 3.21, and it is tight by 3.23. The link of each vertex is homeomorphic to S^5. Therefore 2^K carries the structure of a compact topological manifold. On the other hand the polyhedral decomposition of 2^K as a subcomplex of C^d does not induce a PL structure since the link of a certain 2-face is the triangulated Σ^3.

4. $(k-1)$-connected $2k$-manifolds

For a polyhedral embedding of a $(k-1)$-connected $2k$-manifold M the tightness is equivalent to a generalized Two-piece-property saying that every hyperplane cuts M into at most two pieces such that each piece is again $(k-1)$-connected, compare 3.1. If a polyhedral $(k-1)$-connected $2k$-manifold is a subcomplex of the boundary complex of a convex polytope, then by 3.8 the tightness of the embedding into the ambient space just means that the manifold contains the k-dimensional skeleton of the polytope. This makes it possible to investigate higher dimensional analogues of 2.16, 2.17 and 2.21. Generalized Heawood inequalities are obtained between the dimension of the polytope and the Euler characteristic of the manifold. Fortunately, in this case the Euler characteristic $\chi(M) = 2 + (-1)^k b_k(M)$ contains an essential information about the topology of M, and in some sense $b_k(M)$ counts the 'genus' of the manifold. The case of equality in the generalized Heawood inequalities corresponds to the case of tight triangulations of such manifolds. Particular examples exist in dimensions 4 and 8; they are discussed in Section 4B and Section 4C. These results suggest that the minimum number of vertices for any triangulation of a $(k-1)$-connected $2k$-manifold is essentially the maximum dimension for tight polyhedral embeddings of this manifold. This and other conjectures and a number of open problems are mentioned.

4A. Generalized Heawood inequalities

4.1 Proposition: *Assume that $M \subset P \subset E^d$ is a subcomplex of a convex d-polytope P (such that $P = \mathcal{H}(M)$) and assume that the underlying set of M is homeomorphic to a $(k-1)$-connected $2k$-manifold. Then (i) and (ii) are equivalent:*

    (i) M is tight in E^d.

    (ii) M is k-Hamiltonian in P.

In the particular case of a combinatorial manifold (iii) and (iv) are equivalent:

    (iii) The triangulation is tight.

    (iv) The triangulation is $(k+1)$-neighborly.

This follows directly from 3.8 and 3.18, or 3.10 and 3.18, respectively.
Note that for $k \geq 2$ the tightness of a $(k-1)$-connected $2k$-manifold does not depend on the choice of the field F because there is no torsion in the kth homology. Condition (iv) just means that $S_k(\Delta^d) \subset M \subset Sk_{2k}(\Delta^d)$. According to [Dc] the $(k+1)$-skeleton determines the triangulation of a $2k$-manifold completely. In particular, any $(k+2)$-neighborly combinatorial $2k$-manifold is a sphere, triangulated as the boundary complex of a $(2k+1)$-simplex. This follows from 3.18 and 3.6.

4.2 Theorem ([Kü-Sch]): *For arbitrary given numbers k, d satisfying $d \geq 2k+1$ there is a tight and substantial polyhedral embedding of a $(k-1)$-connected $2k$-manifold into E^d. Particular examples are PL homeomorphic to a connected sum of copies of $S^k \times S^k$.*

PROOF: Let K be a k-neighborly triangulation of S^{2k-1} with d vertices (see 3.9) and define $M := 2^K \subset C^d \subset E^d$. Then M is $(k-1)$-connected and tight by 3.8 and 3.18, or by 4.1. In the particular case of the cyclic polytope $K = \partial C(d, 2k)$, 2^K is PL homeomorphic to a connected sum of $(-1)^k \frac{1}{2}(\chi(d,k)-2)$ copies of $S^k \times S^k$ where

$$\chi(d,k) := 2 \cdot \chi\left(Sk_k(C^{d-k-1})\right)$$

is the Euler characteristic of every k-Hamiltonian submanifold of C^d. For the details of the proof see [Kü-Sch]. It was shown there that 2^K is PL homeomorphic to the boundary of a 'thickening' of $Sk_k(C^{d-k-1})$.

For any such K the intersection form of 2^K on $H_k(2^K)$ is even and can be decomposed into a sum of copies of

$$\begin{pmatrix} 0 & 1 \\ (-1)^k & 0 \end{pmatrix}$$

which puts strong restrictions on the possible PL topological types of 2^K, see [W]. It does not seem to be known whether or not there are distinct topological types of k-Hamiltonian submanifolds of C^d for $k \geq 2$. See 4.15 for an example with odd intersection form as a subcomplex of a truncated cube. For $(k-1)$-connected $2k$-manifolds as tight slicings of $(2k+1)$-pseudomanifolds compare 7.17.

4.3 Proposition (Dehn-Sommerville equations): *Let f_i denote the number of i-dimensional simplices in a simplical complex where formally $f_{-1} := 1$. Then in terms of the f-vector $(f_{-1}, f_0, f_1, \ldots, f_{d-1})$ for any $(d-1)$-dimensional com-binatorial manifold $\underline{M}$ (in particular for the boundary complex of a simplicial d-polytope) the following <u>Dehn-Sommerville equations</u> hold:*

$$\sum_{i=0}^{d-1}(-1)^i f_i = \chi(M) \qquad\qquad (= 0 \text{ if } d \text{ is even})$$

$$\sum_{i=2j-1}^{d-1}(-1)^i \binom{i+1}{2j-1} f_i = 0 \quad \text{for } 1 \le j \le (d-1)/2 \quad \text{if } d \text{ is odd}$$

$$\sum_{i=2j}^{d-1}(-1)^i \binom{i+1}{2j} f_i = 0 \qquad\quad \text{for } 1 \le j \le (d-2)/2 \quad \text{if } d \text{ is even.}$$

In terms of the $\underline{\text{h-vector}}$ $(h_0, h_1, \ldots h_d)$, defined by

$$h_j = \sum_{i=-1}^{j-1}(-1)^{j-i-1}\binom{d-i-1}{j-i-1} f_i,$$

in particular $h_d = (-1)^d(1 - \chi(M))$, these equations read as follows:

If $d = 2k+1$:

$$h_j - h_{d-j} = (-1)^{d-j}\binom{d}{j}(\chi(M) - 2) \quad \text{for } 0 \le j \le k.$$

$$(= 0 \text{ for a polytope})$$

If $d = 2k$:

$$h_j - h_{d-j} = 0 \quad \text{for } 0 \le j \le k-1.$$

SKETCH OF PROOF: The first equation of the Dehn-Sommerville equations is just the Euler equation. We obtain the other equations as sums of the Euler equations of all the links of $2j$-simplices if d is odd or of the $(2j-1)$-simplices if d is even. These links are odd-dimensional spheres. In fact all these equations hold for $(d-1)$-dimensional Euler manifolds. Usually these equations are formulated only for the case of simplicial polytopes, see [Kl1], [Grü], [MM-Sh], [Rn2]. In the general case the only difference is that we have to use $\chi(M)$ instead of $\chi(S^{d-1})$. G.Kalai [Kal1] suggested to modify the definition of the h-vector by the Betti numbers and to replace h_j by $h_j + \binom{d}{j}\sum_{i=1}^{j-1}(-1)^i b_{j-i}(M)$. Then the Dehn-Sommerville equations for a combinatorial manifold read as $h_j - h_{d-j} = 0$ in any case.

Similar equations hold if M is built up by cubes rather than simplices, see [Grü;9.4].

Without a proof we recall the following important result about simplicial polytopes:

4.4 Theorem (Generalized Lower Bound Theorem): *Let P be a simplicial d-polytope. Then for $0 \leq j \leq (d-1)/2$ the following inequality holds:*

$$h_{j+1} - h_j \geq 0$$

or, equivalently,

$$\sum_{i=-1}^{j} (-1)^{j-i} \binom{d-i}{j-i} f_i \geq 0.$$

In the case $j = 0$ this is just the trivial inequality $f_0 \geq d+1$. For $j = 1$ it coincides with D. Barnette's Lower Bound Theorem [Bar1] $f_1 \geq d \cdot f_0 - \binom{d+1}{2}$, for $j \geq 2$ it is different.

4.4 was conjectured by P. McMullen and D. Walkup [MM-Wa] and proved by R. Stanley [Sta2], [Sta3], for an independent and more geometric proof see [MM]. It is part of the so-called 'g-theorem' characterizing the possible f-vectors of simplicial polytopes, compare [Kal3;1.5].

4.5 Theorem ([Kü8]): *Let P be a simplicial d-polytope and let $M \subset P \subset E^d$ be a tight subcomplex which contains all vertices of P and which is a $(k-1)$-connected combinatorial $2k$-manifold. Then the following holds:*

$$\binom{d-k-1}{k+1} \leq (-1)^k \binom{2k+1}{k+1} (\chi(M) - 2) = \binom{2k+1}{k+1} b_k(M).$$

Moreover, for $d \geq 2k + 2$ equality holds if and only if P is a simplex (and, consequently, if M is a tight triangulation).

REMARK: Note that in the case $d = 2k + 1$ the inequality in 4.5 holds trivially, but no conclusion about the case of equality is possible since the boundary of any d-polytope is an example. It is sufficient to assume that M is an Euler manifold with vanishing homology $H_i(M)$ for $i = 1, \ldots, k-1$. The inequality above may be considered as a kind of a 'generalized Heawood inequality', compare 2.17. It seems that for fixed d the right hand side of this inequality gives the minimal 'genus' of a $2k$-manifold admitting an embedding of the complete k-complex with $d+1$ vertices, compare the theorem of van Kampen and Flores for the case of genus zero:

THEOREM (E. van Kampen and A. Flores, [Grü;11.2]): *There is no topological embedding of the underlying set of $Sk_n(\triangle^{2n+2})$ into E^{2n} or S^{2n}.*

An analogue for the complex projective plane would say that there is no topological embedding of the underlying set of $Sk_2(\triangle^9)$ into $\mathbb{C}P^2$ because the generalized

Heawood inequality is not satisfied in this case. Compare 4.14 for an embedding $Sk_2(\Delta^8) \to \mathbf{C}P^2$.

A reformulation of 4.5 is the following:

COROLLARY: *Let P be a simplicial d-polytope and let $M \subset P$ be a k-Hamiltonian $2k$-dimensional Euler manifold where $d \geq 2k+2$. Then the following holds:*

$$\binom{d-k-1}{k+1} \leq (-1)^k \binom{2k+1}{k+1} (\chi(M) - 2)$$

with equality if and only if P is a simplex.

4.6 Proposition: *Assume that M is a $(k-2)$-tight triangulation of a $(k-2)$-connected $2k$-manifold with n vertices. Then the following holds:*

$$\binom{n-k-2}{k+1} \geq (-1)^k \binom{2k+1}{k+1} (\chi(M) - 2)$$

with equality if and only if M is $(k-1)$-connected and the triangulation is tight.

4.7 Corollary: *An n-vertex triangulation of a $(k-1)$-connected $2k$-manifold is tight if and only if*

$$\binom{n-k-2}{k+1} = (-1)^k \binom{2k+1}{k+1} (\chi(M) - 2)$$

Note that 4.7 follows as a corollary of 4.5 and of 4.6 independently if $n \geq 2k+3$, compare 2.16 for the case $k = 1$. Particular integer solutions of this equation in 4.7 are

$n = 2k+2, \quad \chi = 2 \quad$ (boundary of a $(2k+1)$-simplex)

$n = 3k+3, \quad \chi = 2 + (-1)^k$ (this implies $k = 1, 2, 4$ or 8, see [EK],[Br-Kü3])

$n = 3k+4, \quad \chi = 2 + 2(-1)^k$.

4.5, 4.6, and 4.7 correspond more or less to 2.17, 2.21, and 2.16 in the case $k = 1$. However, in 2.17 it is neither assumed that M is a subcomplex of P nor that P is simplicial. We conjecture that these technical assumptions are not really necessary:

Conjecture A: *4.5 holds for any tight polyhedral embedding of a $(k-1)$-connected $2k$-manifold into E^d. More precisely, the inequality in 4.5 holds for any such embedding with equality only for subcomplexes of the d-simplex, if $d \geq 2k+2$.*

For a weaker version one may start with a tight subcomplex of a convex polytope which is a $(k-1)$-connected $2k$-manifold.

Conjecture B: *4.6 holds for any combinatorial $2k$-manifold with n vertices.*

For a weaker version one would consider only $(k-1)$-connected $2k$-manifolds.

Conjecture A is true for $k=1$ by 2.17. Conjecture B is true for several cases, see 4.8. It is a weak version of an Upper Bound Conjecture for combinatorial manifolds, compare [Kl2]. The classical Upper Bound Theorem for simplicial spheres, proved by R. Stanley [Sta1], states that $h_i \leq \binom{n-d+i-1}{i}$ for $0 \leq i \leq (d-1)/2$ if d is odd and for $0 \leq i \leq (d-2)/2$ is d is even. Conjecture B states that the inequality $h_{k+1}(M) - h_k(M) \leq \binom{n-k-2}{k+1}$ holds for any combinatorial $2k$-manifold with n vertices.

PROOF OF 4.5: In order to illustrate the method let us first give the proof in case $k=1$. By assumption P coincides with the convex hull of M and the numbers of vertices and edges of P and M coincide. However, the h-vectors are different. We denote them by $h_i(P)$ and $h_i(M)$, respectively. Using 4.4 for $j=0$ and $j=1$ we obtain:

$$3(2-\chi(M)) = h_2(M) - h_1(M) = f_1 - 3f_0 + 6$$
$$= h_2(P) - h_1(P) + (d-3)f_0 - \binom{d+1}{2} + 6$$
$$\geq (d-3)f_0 - \binom{d+1}{2} + 6$$
$$= (d-3)(h_1(P) - h_0(P)) + (d-3)(d+1) - \binom{d+1}{2} + 6$$
$$\geq (d-3)(d+1) - \binom{d+1}{2} + 6$$
$$= \binom{d-2}{2}$$

where equality implies $f_0 = d+1$ provided that $d \geq 4$. Vice versa, if P is a simplex then $h_{j+1}(P) - h_j(P) = 0$ holds for each j.

For arbitrary k the assertion is trivial if $d = 2k + 1$, so we assume $d \geq 2k + 2$. We start with the equation

$$h_{k+1}(M) - h_k(M) = (-1)^k \binom{2k+1}{k+1} (\chi(M) - 2)$$

from 4.3 and put in successively the inequalities of 4.4 for $j = k, k-1, \ldots, 2, 1, 0$. Each time $h_{j+1}(P) - h_j(P)$ is multiplied by a factor which will be shown to be positive. By 3.5 M contains $Sk_k(P)$, hence the first parts $(f_{-1}, f_0, f_1, \ldots, f_k)$ of the f-vectors of M and P coincide.

Now we obtain the following sequence of inequalities:

$$(-1)^k \binom{2k+1}{k+1} (\chi(M) - 2) \quad = h_{k+1}(M) - h_k(M)$$

$$= \sum_{i=-1}^{k} (-1)^{k-i} \binom{2k+1-i}{k+1} f_i$$

$$\geq \sum_{i=-1}^{k-1} (-1)^{k-i-1} c^i_{k-1,d} f_i$$

$$\cdots$$

$$\geq \sum_{i=-1}^{j} (-1)^{j-i} c^i_{j,d} f_i$$

$$\cdots$$

$$\geq c^1_{1,d} f_1 - c^0_{1,d} f_0 + c^{-1}_{1,d}$$

$$\geq c^0_{0,d} f_0 - c^{-1}_{0,d}$$

$$\geq c^0_{0,d}(d+1) - c^{-1}_{0,d} = c^{-1}_{-1,d}$$

where the $c^i_{j,d}$ are certain coefficients.

For our purpose it is sufficient to compute $c^j_{j,d}$ for each j and to show that it is positive. First of all

$$c^k_{k,d} = 1$$

$$c^{k-1}_{k-1,d} = d - k + 1 - \binom{k+2}{k+1} = d - 2k - 1 > 0.$$

Furthermore the $c^i_{j,d}$ obey the recursion formula

$$c_{k,d}^i = \binom{2k+1-i}{k+1}$$

$$c_{j-1,d}^i = c_{j,d}^i \binom{d-i}{j-i} - c_{j,d}^i \text{ for } j > i.$$

The proof is completed by the following lemma which in particular says

$$c_{0,d}^0 = \binom{d-k-2}{k} > 0 \quad \text{and} \quad c_{-1,d}^{-1} = \binom{d-k-1}{k+1}.$$

LEMMA: $c_{i,d}^i = \binom{d-k-i-2}{k-i}$ for $-1 \leq i \leq k$.

PROOF: The assertion is true for $i = k : c_{k,d}^k = 1$, see above. By the recursion formula we have

$$\begin{aligned}
c_{i,d}^i &= c_{i+1,d}^{i+1} \binom{d-i}{i} - c_{i+1,d}^i \\
&= c_{i+1,d}^{i+1} \binom{d-i}{i} - c_{i+2,d}^{i+2} \binom{d-i}{2} + c_{i+2,d}^i \\
&= \sum_{j=i+1}^{k} (-1)^{j-i-1} c_{j,d}^j \binom{d-i}{j-i} + (-1)^{k-i} \binom{2k+1-i}{k+1}.
\end{aligned}$$

Consequently, if $c_{j,2k+2}^j = 1$ for $j \geq i+1$ then it follows

$$\begin{aligned}
c_{i,2k+2}^i &= \sum_{j=i+1}^{k} (-1)^{j-i-1} \binom{2k+2-i}{j-i} + (-1)^{k-i} \binom{2k+1-i}{k+1} \\
&= 1 - (-1)^{k-i} \binom{2k+1-i}{k-i} + (-1)^{k-i} \binom{2k+1-i}{k+1} \\
&= 1.
\end{aligned}$$

Inductively we get $c_{i,2k+2}^i = 1$ for $-1 \leq i \leq k$.

Now we compare the numbers $c_{i,d}^i$ with the entries of Pascal's triangle. We shall prove <u>Pascal's rule</u>

(PR) $$c_{i,d}^i + c_{i+1,d+1}^{i+1} = c_{i,d+1}^i$$

by induction backwards. (PR) holds for $i = k - 1$ (see above). Now assume that (PR) holds for $j = i + 1, i + 2, \ldots, k - 1$. This implies

$$
\begin{aligned}
c_{i,d}^i + c_{i+1,d+1}^{i+1} &= \sum_{j=i+1}^{k} (-1)^{j-i-1} c_{j,d}^j \binom{d-i}{j-i} + (-1)^{k-i} \binom{2k+1-i}{k+1} + \\
&\quad + \sum_{j=i+2}^{k} (-1)^{j-i} c_{j,d+1}^j \binom{d-i}{j-i-1} + (-1)^{k-i-1} \binom{2k-i}{k+1} \\
&= \sum_{j=i+1}^{k-1} (-1)^{j-i-1} \left(c_{j,d}^j + c_{j+1,d+1}^{j+1} \right) \binom{d-i}{j-i} + \\
&\quad + (-1)^{k-i-1} c_{k,d}^k \binom{d-i}{k-i} + (-1)^{k-i} \binom{2k+1-i}{k+1} + \\
&\quad + (-1)^{k-i-1} \binom{2k-i}{k+1} \\
&= \sum_{j=i+1}^{k} (-1)^{j-i-1} c_{j,d+1}^j \binom{d-i}{j-i} + \\
&\quad + (-1)^{k-i} \binom{2k+1-i}{k+1} + (-1)^{k-i-1} \binom{2k-i}{k+1} \\
&= \sum_{j=i+1}^{k} (-1)^{j-i-1} c_{j,d+1}^j \binom{d+1-i}{j-i} + (-1)^{k-i} \binom{2k+1-i}{k+1} - \\
&\quad - \sum_{j=i+1}^{k} (-1)^{j-i-1} c_{j,d+1}^j \binom{d-i}{j-i-1} + (-1)^{k-i-1} \binom{2k-i}{k+1} \\
&= c_{i,d+1}^i.
\end{aligned}
$$

This completes the proof of 4.5. The coefficients $c_{i,d}^i$ coincide with the entries of Pascal's triangle as follows:

$$i = k \qquad\qquad d = 2k + 2$$

$$
\begin{array}{ccccccccc}
&&&& \searrow & & \swarrow \\
&&&& 1 \\
&&& 1 && 1 \\
&& 1 && 2 && 1 \\
& 1 && 3 && 3 && 1 \\
1 && 4 && 6 && 4 && 1
\end{array}
$$

REMARK: We see again that for $d = 2k + 1$ no conclusion is possible in the case of equality because of $c_{0,2k+1}^0 = 0$. There is also a direct inductive proof for the lemma using the Vandermonde convolution formula, see [Kü8]. However, we need Pascal's rule in the proof of 7.11.

PROOF OF 4.6: By 3.6 the triangulation is k-neighborly:

$$f_0 = n, \quad f_1 = \binom{n}{2}, \ldots, \quad f_{k-1} = \binom{n}{k}.$$

The inequality $f_k \leq \binom{n}{k+1}$ holds trivially. This implies

$$(-1)^k \binom{2k+1}{k+1}(\chi(M) - 2) \ = h_{k+1}(M) - h_k(M)$$

$$= \sum_{i=-1}^{k-1} (-1)^{k-i} \binom{2k+1-i}{k-i}\binom{n}{i+1} + f_k$$

$$\leq \sum_{i=-1}^{k} (-1)^{k-i} \binom{2k+1-i}{k-i}\binom{n}{i+1}$$

$$= \binom{n-k-2}{k+1}$$

where the last equality holds by the Vandermonde convolution formula (see [Rn1;1.3])

$$\binom{n+p}{\ell} = \sum_{i=0}^{\ell} \binom{p}{i}\binom{n}{\ell-i}$$

which is valid for any n, p, ℓ including negative p or n.
The case of equality in the inequality is equivalent to $f_k = \binom{n}{k+1}$. The assertion follows from 3.6 and 3.9.

4.8 Remark: Conjecture B has been made also by G. Kalai who moreover suggested to replace in the right hand side of the inequality the expression $(-1)^k (\chi(M) - 2)$ by a (possibly weighted) sum of the Betti numbers $b_1(M), \ldots, b_{2k-1}(M)$, compare [Kal1], [Bj-K].
Conjecture B is true in each of the following cases:

1. $k = 1$, see 2.21.

2. $k = 2$, see 4.9.

3. M is a 'manifold like a projective plane' in the sense of [EK], see [Br-Kü3].

4. $n \leq 3k + 3$, see [Br-Kü3].

5. M has the same homology as $S^j \times S^{2k-j}$, $j < k$, see [Br-Kü3].

6. $\chi(M) \leq 2$ if k is even, $\chi(M) \geq 2$ if k is odd (trivial).

7. $n > k^2 + 4k + 2$, see below.

Compare a similar discussion in [Grü;10.1] for the Upper Bound Conjecture for polytopes, before the solution was given in [MM-Sh].

In general Conjecture B would follow if one could prove the inequality

$$ f_k - (k+2)f_{k-1} \leq \binom{n}{k+1} - (k+2)\binom{n}{k} $$

with equality if and only if $f_k = \binom{n}{k+1}$. In particular, this holds under the extra assumption $n > k^2 + 4k + 2$ (condition 7 above) as follows:

The inequality $(n-k)f_{k-1} \geq (k+1)f_k$ holds since each $(k-1)$-simplex is contained in at most $(n-k)$ k-simplices. This implies

$$ f_k - (k+2)f_{k-1} \leq f_k \left(1 - \frac{(k+1)(k+2)}{n-k} \right) $$

$$ = f_k \, \frac{n - k^2 - 4k - 2}{n-k} $$

$$ \leq \binom{n}{k+1} \left(1 - \frac{(k+1)(k+2)}{n-k} \right) $$

$$ = \binom{n}{k+1} - (k+2)\binom{n}{k}. $$

The crucial condition is equivalent to the inequality

$$ \binom{n}{k+1} - f_k \geq (k+2) \left(\binom{n}{k} - f_{k-1} \right), $$

i.e., each missing $(k-1)$-simplex should lead to at least $(k+2)$ missing k-simplices. This holds trivially for $k = 1$. It is also true for $k = 2$, see the proof of 4.9.

4B. Triangulated 4-manifolds

The combinatorics of triangulated 4-manifolds with n vertices is governed by the Dehn-Sommerville equations 4.3 as follows

$$
\begin{aligned}
n - f_1 + f_2 - f_3 + f_4 &= \chi(M) \\
2f_1 - 3f_2 + 4f_3 - 5f_4 &= 0 \\
2f_3 - 5f_4 &= 0.
\end{aligned}
$$

If we eliminate f_3 and f_4 from this system of linear equations we obtain the equation $f_2 - 4f_1 + 10n = 10\chi(M)$. This leads to a lower bound for the number of vertices as follows:

4.9 Theorem (W. Kühnel, [Kü5;4.1]): *For any combinatorial 4-manifold M with n vertices the inequality*

$$
\binom{n-4}{3} \geq 10\big(\chi(M) - 2\big)
$$

holds with equality if and only if the triangulation is 3-neighborly or, equivalently, if M is simply connected and the triangulation is tight.

PROOF: If $n \leq 8$ then it follows that M is homeomorphic to S^4 (see [Bar-G], [Br-Kü3]), thus $\chi(M) = 2$, and the assertion is trivial.

Now let $\overline{f}_2 := \binom{n}{3} - f_2$, $\overline{f}_1 := \binom{n}{2} - f_1$, and for each vertex i let $f_0(i)$ and $f_1(i)$ denote the number of vertices and edges in the link of i. Let $\overline{f}_0(i) := n - 1 - f_0(i)$, $\overline{f}_1(i) := \binom{n-2}{2} - f_1(i)$. The link of i is a 3-sphere with at least 5 vertices, therefore $\overline{f}_0(i) \leq n - 6$.

This implies the inequality

$$
\begin{aligned}
\overline{f}_1(i) \quad &\geq f_0(i) \cdot \overline{f}_0(i) + \binom{\overline{f}_0(i)}{2} \\
&= \overline{f}_0(i)\big(n - 1 - \tfrac{1}{2}(\overline{f}_0(i) + 1)\big) \\
&\geq \overline{f}_0(i)\big(n - 1 - \tfrac{1}{2}(n - 5)\big) \\
&= \overline{f}_0(i) \cdot \tfrac{1}{2}(n + 3) \\
&\geq \overline{f}_0(i) \cdot 6
\end{aligned}
$$

whenever $n \geq 9$.

Summation over all i gives

$$3\overline{f}_2 \geq 2 \cdot 6 \cdot \overline{f}_1$$

with equality if and only if $\overline{f}_2 = \overline{f}_1 = 0$. Compare [Grü;10.1] for the same inequality in the context of the classical Upper Bound Conjecture.

This inequality $\overline{f}_2 \geq 4\overline{f}_1$ implies

$$10\big(\chi(M) - 2\big) \;=\; h_3(M) - h_2(M)$$

$$= f_2 - 4f_1 + 10n - 20$$

$$\leq \binom{n}{3} - 4\binom{n}{2} + 10n - 20$$

$$= \binom{n-4}{3}$$

with equality if and only if $f_2 = \binom{n}{3}$.

Although we may consider 4.9 as a lower bound for the number of vertices, it is more of the type 'Upper Bound Theorem' because $\binom{n-4}{3}$ appears as an upper bound for $h_3(M) - h_2(M)$.

4.10 Corollary: *Every tight triangulation of a simply connected 4-manifold has the minimal possible number of vertices among all combinatorial 4-manifolds with the same Euler characteristic.*

Conjecture C: *Every tight triangulation has the minimal possible number of vertices among all combinatorial manifolds of the same (PL) topological type.*

Conjecture C is true for 2-manifolds and for simply connected 4-manifolds and for several other cases, see Chapter 5 and Chapter 7.

4.11 Corollary: *The number of vertices of any combinatorial triangulation is at least*

(i) *9 for* CP^2

(ii) *10 for* $CP^2 \sharp CP^2$, $CP^2 \sharp (-CP^2)$, $S^2 \times S^2$

(iii) *13 for a regular cubic surface in* CP^3

(iv) 14 *for a connected sum of six copies of* $S^2 \times S^2$

(v) 16 *for a regular quartic surfaces in* $\mathbf{C}P^3$ *(or any K3-surface)*

(vi) 24 *for a regular sextic surface in* $\mathbf{C}P^3$,

and in the cases (i), (ii), (iv), (v) such a triangulation with 9, 10, 14, *or* 16 *vertices is necessarily tight.*
However, in case (ii) this bound is not attained, see 4.12.

QUESTION: *Does there exist a tight 16-vertex triangulations of a K3-surface?* There exists a 16-vertex triangulation of the singular 16-nodal Kummer variety which has the minimal number of vertices but which is not tight, see [Kü4]. Compare [Tr] for a cell decomposition of a regular quartic surface (the Fermat surface) with 16 vertices. However, it is not obvious how to make this into a triangulation with 16 vertices. Recently M. Casella came up with a very interesting 16-vertex triangulation of a simply connected 4-manifold satisfying equality in 4.9. It is certainly a tight triangulation but its topological type is not yet determined.

4.12 Theorem ([Kü-L1]): *There is exactly one (up to relabelling) tight triangulation of a simply connected 4-manifold with* $n \leq 13$ *vertices which is topologically distinct from the sphere. This has* $n = 9$ *vertices and* $\chi = 3$.

In particular there is no tight triangulation of any of the manifolds $\mathbf{C}P^2 \sharp \mathbf{C}P^2$, $\mathbf{C}^2 \sharp (-\mathbf{C}P^2)$ or $S^2 \times S^2$. The assumption $n \leq 13$ can be replaced by $\chi \leq 13$.

PROOF: By 4.7 the number n of vertices of such a tight triangulation satisfies

$$\binom{n-4}{3} = 10\big(\chi(M) - 2\big).$$

This implies $n \equiv 0, 1, 4, 5, 6, 9, 10, 14, 16 \bmod 20$. Therefore for $n \leq 13$ there are only the cases $n = 6$ (the boundary of the 5-simplex), $n = 9$ and $n = 10$. The link of each vertex in such a combinatorial 4-manifold with 9 or 10 vertices must be a 2-neighborly combinatorial 3-sphere with 8 or 9 vertices, respectively. These are classified, and there are 4 different types with 8 vertices and 50 ones with 9 vertices, see [Grü-S], [AS1], [Bar2]. The proof given in [Kü-L1] consists of a computer-aided enumeration in each of these cases. It turned out that only one type of a link is appropriate, namely the Brückner-Grünbaum sphere with 8 vertices, called $\mathcal{M}$ in [Grü-S]. For $n = 9$ the uniqueness has been shown independently in [AM] and [BD]. The lengthy details of this combinatorial uniqueness are omitted here. An explicit description of the example is given in 4.13.

4.13 Theorem ([Kü-Ba], [MY]): *There exists a unique tight 9-vertex triangulation of the complex projective plane* $\mathbf{C}P^2$.

PROOF: As a necessary condition the f-vector of such a triangulation can be computed from 4.3:

$$f_0 = 9, \quad f_1 = \binom{9}{2} = 36, \quad f_2 = \binom{9}{3} = 84, \quad f_3 = 90, \quad f_4 = 36.$$

To construct the triangulation, called $\mathbf{C}P_9^2$, we denote the nine vertices by 1, 2, 3, ..., 9 and we take the union of the two orbits of the 4-dimensional simplices

$$\langle 12456 \rangle \text{ and } \langle 12459 \rangle$$

under the action of a group H_{54} on $\{1, 2, \ldots, 9\}$ which is generated by

$$\alpha = (147)(258)(369)$$
$$\beta = (123)(465)$$
$$\gamma = (12)(45)(78).$$

This group is a 2-fold extension of the Heisenberg group over $\mathbf{Z}_3$ or of the Burnside group $B_{2,3}$ with two generators of order 3, see [Cox-M;6.8]. γ corresponds to the action of the complex conjugation, in fact its fixed point set is combinatorially isomorphic to an $\mathbf{R}P_6^2$.

The complete list of all the 36 4-dimensional simplices is the following:

12456	45789	78123
23564	56897	89231
31645	64978	97312

12459	45783	78126	23649	56973	89316	31569	64893	97236
23567	56891	89234	31457	64781	97124	12647	45971	78314
31648	64972	97315	12568	45892	78235	23458	56782	89125

The triangulation is unique by 4.12 (up to relabelling of the vertices). The link of each vertex is combinatorially isomorphic to the Brückner-Grünbaum sphere denoted by $\mathcal{M}$ in [Grü-S]. This is a peculiar triangulation of S^3. It is not the boundary complex of any convex 4-polytope but it admits a simplexwise linear embedding as a convex polyhedron in E^4 which induces a so-called flattening of each vertex star, compare [Mi]. Therefore $\mathbf{C}P_9^2$ has the structure of a PL manifold, and in particular it is a combinatorial 4-manifold with $\chi = 3$. The tightness of the triangulation holds by 3.10 and 3.18, it also follows that the underlying

manifold is simply connected. This implies that it is homotopy equivalent to
CP^2 by classical results in topology, see [Wht]. By M. Freedman's classification
of simply connected 4-manifolds it must be homeomorphic to CP^2, see [Ki]. On
the other hand several direct proofs are available to show that it is PL homeomorphic to CP^2, compare [Kü5;4.4]. This involves a decomposition into three
PL 4-balls induced by the moment map into E^2 [Kü-Ba], or a complex crystal in
$E^4 \cong C^2$ admitting a branched simplicial covering onto CP_9^2 [MY], or a Dirichlet
tessellation in the Fubini-Study metric of CP^2 which is the combinatorial dual
of CP^2 [Br3].

4.14 Corollary: *There exists a tight and substantial polyhedral embedding*
$CP^2 \to E^8$. *Any linear projection in general position induces a tight embed-*
ding into E^7.

PROOF: The canonical embedding of CP_9^2 into an 8-simplex $\triangle^8$ is tight by 4.7:

$$Sk_2(\triangle^8) \subset CP_9^2 \subset Sk_4(\triangle^8) \subset E^8.$$

Any linear projection preserves the tightness by 3.16, and CP_9^2 is still embedded
in E^7 provided that the 9 vertices are in general position. This follows from the
fact that any two 4-simplices of CP_9^2 together cover at most 8 vertices. This
is a consequence of the remarkable complementarity condition of CP_9^2: *For an*
arbitrary subset W of the set V of vertices either W or $V\backslash W$ spans a simplex of
the triangulation. In particular, opposite to a 4-simplex there is the boundary
of a 3-simplex:

$$(5 \text{ vertices}) \qquad \triangle^4 \longleftrightarrow \partial\triangle^3 \qquad (4 \text{ vertices}),$$

compare the numerical complementarity $f_3 + f_4 = \binom{9}{4}$.
The deformation retract from $CP_9^2 \backslash \triangle^4$ onto $\partial\triangle^3$ is a combinatorial analogue
of the Hopf fibration, regarded on the Hopf 2-disc bundle over S^2. In the link
of each vertex the complementarity condition implies that there is no pair of
disjoint tetrahedra. In fact, the Brückner-Grünbaum sphere $\mathcal{M}$ is the only 2-
neighborly 3-sphere with 8 vertices which has this property [Grü-S], [Bar2].

Compare [Br-Kü4] or [AM] for a further discussion of this complementarity condition. Somehow it corresponds to the classical duality (*point* $\leftrightarrow$ *line*) in a projective plane. Compare also the tight algebraic standard embedding (or Veronese
embedding)

$$CP^2 \to S^7 \subset E^8$$

which projects down to an embedding into E^7 whenever it becomes an immersion, because any secant is parallel to a tangent, see [Ma], [Kui4].

QUESTION (N.H. Kuiper, [Kui9]): *Are there tight and substantial topological embeddings* $CP^2 \to E^8$ *which are distinct from the algebraic standard embedding and from the canonical embedding of* CP_9^2 *(up to projective transformations)?*

By 4.5 there is no one in the boundary complex of a simplicial polytope. Conjecture A would imply the uniqueness in the polyhedral case.

REMARK: The 36 edges of CP_9^2 split into two H_{54}-orbits. By a suitable choice of the lengths of these edges it is possible to find an abstract H_{54}-invariant PL metric on CP^2 which has nonnegative curvature in the following sense: at each triangle (= codimension-2-face) the sum of the interior dihedral angles of the adjacent 4-simplices is not greater than 2π. In fact the same metric is induced by the (flat) complex crystal described in [MY]: The curvature is zero everywhere except along the branch locus of a covering from the 4-torus onto CP^2. In the triangulation this is just the H_{54}-orbit of the triangle $\langle 169 \rangle$ which is a 2-torus with 9 vertices and 18 triangles. By a theorem of J. Cheeger [Ch] a positive polyhedral curvature of this kind on a simply connected 4-manifold M is possible only if M is a homology sphere. Similar considerations about the curvature of invariant PL metrics on CP_9^2 have been made by J. Hartle [Har] in order to determine discrete solutions of Einstein's equations.
The combinatorial formula for the first Pontrjagin number of a 4-manifold (compare [LR], [MP]) has been explicitly evaluated for CP_9^2 by L. Milin [Mi].

4.15 Example: *For any* m, $1 \le m \le 256$, *there is a tight polyhedral embedding into* E^8 *of a simply connected 4-manifold with* $\mathrm{rank}(H_2) = 62 + m$ *whose intersection form on* H_2 *is odd.*

Construction: We consider $2^{\mathcal{M}}$ where $\mathcal{M}$ is the Brückner-Grünbaum sphere mentioned in 4.12 and 4.13. $2^{\mathcal{M}}$ is a 2-Hamiltonian 4-dimensional subcomplex of C^8 which is a simply connected manifold and which is tight in E^8 by 4.1 or by 3.24. Its Euler characteristic is

$$\chi(8,2) = 2\chi\big(Sk_2(C^5)\big) = 64 \quad (\text{compare 4.2}).$$

Its intersection form on the 62-dimensional second homology splits into 31 direct summands $\binom{0\,1}{1\,0}$ because each missing triangle in $\mathcal{M}$ bounds a polyhedral 2-disc in $Sk_2(\mathcal{M})$, see [KS;Thm.2]. The link of each vertex is combinatorially equivalent to $\mathcal{M}$. Therefore we can cut off each vertex by a hyperplane section and attach in this hyperplane put a copy of CP_9^2 minus an open vertex star. The resulting manifold is 2-Hamiltonian in the truncated 8-cube and therefore tight by 4.1. This procedure can be repeated m times for each of the $2^8 = 256$ vertices of C^8. The intersection form of this manifold is the one of $2^{\mathcal{M}}$ plus m direct summands (± 1).

4.16 Proposition ([Ba-Kü]): *There is a tight polyhedral embedding* $CP^2 \to E^7$ *which is essentially different from a linear projection of the one in 4.14. It is a simplexwise linear embedding of a 10-vertex triangulation, denoted by* CP_{10}^2.

PROOF: The triangulation is based on the decomposition of the complex projective plane into three 4-balls as 'zones of influence' of three particular points $X = [1,0,0], Y = [0,1,0]$ and $Z = [0,0,1]$, given in homogeneous coordinates. The *equilibrium torus* is the set of points $[z_0, z_1, z_2]$ with the same absolute value of each coordinate z_i. We take the 7-vertex triangulation of this equilibrium torus and then introduce X, Y, Z as extra vertices. Each of the three 4-balls is triangulated as a cone over the boundary complex of the cyclic polytope $C(7,4)$, which occurs in three different - combinatorially equivalent - versions. The complete list of all 42 top-dimensional simplices of CP_{10}^2 is the following:

$$
\begin{array}{cccccc}
0134X & 0123X & 0123Y & 0246Y & 0246Z & 0134Z \\
1245X & 1234X & 1234Y & 1350Y & 1350Z & 1245Z \\
2356X & 2345X & 2345Y & 2461Y & 2461Z & 2356Z \\
3460X & 3456X & 3456Y & 3502Y & 3502Z & 3460Z \\
4501X & 4560X & 4560Y & 4613Y & 4613Z & 4501Z \\
5612X & 5601X & 5601Y & 5024Y & 5024Z & 5612Z \\
6023X & 6012X & 6012Y & 6135Y & 6135Z & 6023Z
\end{array}
$$

The link of X is nothing but the boundary complex of the cyclic polytope $C(7,4)$, each column separately is a 7-vertex solid torus whose boundary is the unique 7-vertex torus. The automorphism group of CP_{10}^2 is the same as the automorphism group of the 7-vertex torus, generated by $T = (0123456)$ and $R = (132645)(XZY)$. It is the group of all affine transformations of the field with 7 elements. T corresponds to the 'translation' $x \mapsto x + 1$, R is a 'rotation' $x \mapsto 3x$.

The embedding into E^7 is defined as follows: We start with the canonical embedding of the 7-vertex torus into $\triangle^6$. Then we regard this 6-space as a hyperplane of E^7 and we put X somewhere outside this hyperplane and Y exactly on the opposite side such that the straight line determined by X and Y meets $\triangle^6$ at an interior point. Finally the vertex Z gets its position at the centre of the 6-simplex. This determine the positions of all the simplices. The embedding contains the 2-skeleton of its convex hull and the interior vertex Z does not produce critical points of index 1 for any height function in general position. It follows that the intersection of the image with any half space is simply connected and that the embedding is tight.

4C. Tight triangulations of manifolds like projective planes

In the case of <u>2-connected 6-manifolds</u> ($k = 3$) no examples are known of tight triangulations except for the boundary of a 7-simplex with $n = 8$, $\chi = 2$. The next integer solution of 4.7 would be $n = 12$, $\chi = 1$ but such a manifold would have to be a 6-dimensional 'manifold like a projective plane' in the sense of [EK] which does not exist. The subsequent case $n = 13$, $\chi = 0$ is still open.

QUESTION: *Does there exist a tight 13-vertex triangulation of $S^3 \times S^3$?*

According to a computer search by U. Brehm and G. Lassmann there is no one with a cyclic automorphism group of order 13.

In the case of <u>3-connected 8-manifolds</u> ($k = 4$) the first example is the boundary of the 9-simplex with $n = 10$, $\chi = 2$. The next integer solutions of 4.7 are

$$n = 15, \quad \chi = 3$$
$$n = 16, \quad \chi = 4$$
$$n = 24, \quad \chi = 70$$

Although nothing seems to be known about the cases $n = 16$ and $n = 24$, there are examples in the case $n = 15$ according to the following theorem:

4.17 Theorem (U. Brehm and W. Kühnel, [Br-Kü4]): *There are at least three combinatorially distinct tight 15-vertex triangulations of an 8-manifold 'like the quaternionic projective plane'.*

4.18 Corollary: *There are at least three tight polyhedral submanifolds of E^{14} which are PL homeomorphic to one 8-manifold 'like the quaternionic projective plane' and which are not affinely equivalent to each other.*

Compare the tight algebraic standard embedding $\mathbf{H}P^2 \to S^{13} \subset E^{14}$, [Kui8].

PROOF OF 4.17: As a necessary condition 4.3 together with $\chi = 3$ implies the following f-vector:

$$f_0 = n = 15,$$

$$f_1 = \binom{15}{2} = 105,$$

$$f_2 = \binom{15}{3} = 455,$$

$$f_3 = \binom{15}{4} = 1365,$$

$$f_4 = \binom{15}{5} = 3003,$$

$$f_5 = 4515,$$

$$f_6 = 4230,$$

$$f_7 = 2205,$$

$$f_8 = 490$$

satisfying the numerical complementarity

$$f_5 + f_8 = 5005 = \binom{15}{6}, \quad f_6 + f_7 = 6435 = \binom{15}{7}.$$

To construct the most symmetric one among the three triangulations, denoted by M_{15}^8, we consider the permutation group acting on $\{1, 2, \ldots, 15\}$ which is generated by P and S:

$$P := (1\,2\,3\,4\,5)(6\,7\,8\,9\,10)(11\,12\,13\,14\,15)$$
$$S := (1\,6\,11)(2\,15\,14)(3\,13\,8)(4\,7\,5)(9\,12\,10).$$

The relations

$$P^5 = S^3 = (PS)^2 = E$$

show that this group is isomorphic to the alternating group A_5 of oder 60, see [Cox-M; Table 5]. The same group action occurs for a polyhedral embedding of the quaternion space into E^4, based on an 8-fold quotient of the 600-cell, see [Br-Kü2].

Now the triangulation M_{15}^8 is defined as the union of the A_5-orbits of the following fourteen 8-simplices A, B, C, ..., N, each given by a 9-tuple of vertices in $\{1, 2, \ldots, 15\}$:

A : ⟨ 1   2   3   6   8   11   13   14   15 ⟩

B : ⟨ 1   3   6   8   9   10   11   12   13 ⟩

C : ⟨ 1   2   6   9   10   11   12   14   15 ⟩

D : ⟨ 1   2   3   4   7   9   12   14   15 ⟩

E : ⟨ 1   2   4   7   9   10   12   13   14 ⟩

F : ⟨ 1   2   6   8   9   10   11   14   15 ⟩

G : ⟨ 1   2   3   4   5   6   9   11   13 ⟩

H : ⟨ 1   3   5   6   8   9   10   11   12 ⟩

I : ⟨ 1   3   5   6   7   8   9   10   11 ⟩

J : ⟨ 1   2   3   4   5   7   10   12   15 ⟩

K : ⟨ 1   2   3   7   8   10   12   13   14 ⟩

M : ⟨ 2   5   6   7   8   9   10   13   14 ⟩

L : ⟨ 3   4   6   7   11   12   13   14   15 ⟩

N : ⟨ 3   4   6   7   10   12   13   14   15 ⟩

These A_5-orbits are of lengths

60 for F, G, H, I, L,	=	300	together
30 for D, E, J, K,	=	120	together
15 for M, N,	=	30	together
20 for C,	=	20	together
10 for A, B,	=	20	together
	=	490	altogether.

The two other triangulations of the same PL manifold are slight modifications of M_{15}^8 in the A_5-orbits of L and N. These three triangulations are distingished from each other by their automorphism groups. Each of them is tight by 4.7, and each of them satisfies the same complementarity condition as CP_9^2 (compare 4.14).

For the details we refer to [Br-Kü4] containing a proof that these triangulations are combinatorial 8-manifolds, and also containing several arguments to support the strong conjecture that M_{15}^8 is PL homeomorphic to HP^2. A certain Dirichlet tessellation of HP^2 seems to be the combinatorial dual of the triangulated M_{15}^8. The remaining problem here is how to verify this by numerical calculations. Each vertex link in M_{15}^8 turned out to be a non-polytopal 7-sphere, a phenomenon already known from CP_9^2, see 4.13. The fixed point sets of certain subgroups of the automorphism group of M_{15}^8 are combinatorially isomorphic to CP_9^2 or RP_6^2, respectively.

In the case of 7-connected 16-manifolds ($k = 8$) the first example is the boundary of the 17-simplex with $n = 18, \chi = 2$. The next integer solutions of 4.7 are

$$n = 27, \quad \chi = 3$$
$$n = 28, \quad \chi = 4$$
$$n = 62, \quad \chi = 151342.$$

QUESTION: *Does there exist a tight 27-vertex triangulation of a 16-manifold 'like the Cayley-plane'?*

This number 27 is suggested independently by

1. the tight algebraic standard embedding $CaP^2 \to S^{25} \subset E^{26}$, see [Kui4],

2. the lower bounds given in [Br-Kü3],

3. the equation 4.7 for tight triangulations,

4. the results about weak cohomology projective spaces given in [AM].

From 4.3 we can compute the following f-vector for such a triangulation:

$$f_i = \binom{27}{i+1} \quad \text{for } 0 \le i \le 8$$

$$
\begin{array}{rcl}
f_9 & = & 8\,335\,899 \\
f_{10} & = & 12\,184\,614 \\
f_{11} & = & 14\,074\,164 \\
f_{12} & = & 12\,301\,200 \\
f_{13} & = & 7\,757\,100 \\
f_{14} & = & 3\,309\,696 \\
f_{15} & = & 853\,281 \\
f_{16} & = & 100\,386
\end{array}
$$

satisfying the numerical complementarity

$$f_j + f_{25-j} = \binom{27}{j+1} \quad \text{for } 9 \le j \le 12.$$

5. 3-manifolds and twisted sphere bundles

For odd-dimensional manifolds it seems to be difficult to transform the tightness of a polyhedral embedding into a simple combinatorial condition, even in the case of tight triangulations. According to 2.2 the 0-tightness of $M \subset E^d$ implies that the complete 1-skeleton of the convex hull must be contained in M. On the other hand 0-tightness is a rather weak condition for 3-manifolds, compare 3.12. If a 3-manifold $M \subset E^d$ contains the complete 2-skeleton of its convex hull $\mathcal{H}$ then $M \cap \partial \mathcal{H}$ is simply connected. This condition is extremely restrictive for 3-manifolds. In particular any 3-neighborly triangulation of a 3-manifold must be $\partial \triangle^4$ with 5 vertices. This follows from the Dehn-Sommerville equations 4.3:

$$
\begin{aligned}
f_0 - f_1 + f_2 - f_3 &= 0 \\
2f_2 - 4f_3 &= 0
\end{aligned}
$$

If we assume $f_0 = n, f_1 = \binom{n}{2}, f_2 = \binom{n}{3}$ then we obtain easily $\binom{n-4}{2} = 0$ or, equivalently, $n \leq 5$. This follows also from 3.6 because such a manifold is a homology sphere by Poincaré duality. In general, the Dehn-Sommerville equations tell us that two combinatorial 3-manifolds with the same numbers of vertices and edges have the same f-vectors. This type of phenomenon is common to all odd-dimensional manifolds, and it makes it difficult to recognize tight triangulations. In order to prove results about tight triangulations of 3-manifolds, one has to look at the top-sets of the canonical embedding in more detail. In particular, the various homological types of these top-sets play an essential role.

Examples of tightly embedded 3-manifolds with high substantial codimension can be constructed as cartesian products of tight surfaces with a convex planar polygon, see 3.16. Other examples can be constructed by the method of Section 3C, see the following Example 5.2.

5A. Tight triangulations of 3-manifolds

5.1 Definition: A simplicial d-polytope P is called <u>stacked</u> if there is a sequence of $P_1, \ldots, P_k$ of simplicial d-polytopes in such a way that P_1 is a simplex, $P_k = P$ and each P_{j+1} can be constructed from P_j by attaching a simplex along a facet of P_j. The boundary complexes of stacked polytopes (so-called <u>stacked spheres</u>) are the only (pseudo-)manifolds satisfying equality in the inequalities of the Lower Bound Theorem [Bar1], [Kal1]. In particular, any triangulated 3-manifold satisfying $f_1 = 4f_0 - 10$ is a stacked sphere.

If a simplicial complex does not contain a certain simplex Δ^k but does contain $\partial\Delta^k$ then Δ^k is called a <u>missing k-face</u> ([Kal1;8.1]) or an <u>empty k-simplex</u> ([Kal3;1.4]).

5.2 Example: *Let K denote the boundary complex of a stacked 3-polytope with d vertices. Then $2^K \subset C^d \subset E^d$ is a tightly embedded polyhedral 3-manifold which is PL homeomorphic to the standard 3-manifold of Heegaard genus $g_{d-1} = 2^{d-4}(d-5) + 1$.*

By a standard 3-manifold of genus g we mean the connected sum of g copies of $S^1 \times S^2$. 2^K is tight in E^d by 3.23, and its topology can be determined by induction on the number of vertices of K, using the numbering of the vertices determined by the stacking. It may also be identified with the boundary of a 4-dimensional thickening of $Sk_1(C^{d-1})$, an orientable 1-handlebody, see [Kü-Sch;Thm.1(iii)].

5.3 Theorem: *Let M be a tight triangulation of a compact 3-manifold with $rk\, H_1(M) \le 1$ which is not homeomorphic to S^3. Then M is the unique 9-vertex triangulation of the 3-dimensional Klein bottle (= the total space of the twisted S^2-bundle over S^1).*

REMARK: This triangulation has been found by D. Walkup [Wa]. Later A. Altshuler and L. Steinberg [AS2] recognized it as the only 9-vertex triangulation of any 3-manifold which is not a sphere. In particular 5.3 says that there is no tight triangulation of $S^1 \times S^2$ or of $\mathbf{R}P^3$ or of any lens space. The minimum number of vertices for any triangulation as well as for a 2-neighborly (or 0-tight) triangulation is

9	for	the 3-dim. Klein bottle	
10	for	$S^1 \times S^2$	[Wa], [Al2], [Al3]
11	for	$\mathbf{R}P^3$	[Wa]
≤ 12	for	$L(3,1)$	(U. Brehm, personal communication).

For the 3-torus there is a triangulation with 15 vertices [Kü-L2] but the minimum number is not known (perhaps 13 ?). Compare [Kü-L3] for 2-neighborly 3-manifolds with a transitive and cyclic or dihedral automorphism group. For explicit triangulations of lens spaces see [Br-Sw].

PROOF: Let n be the number of vertices. Then $f_1 = \binom{n}{2}$ follows from the 0-tightness by 3.10. Hence the Dehn-Sommerville equations imply $f_2 = n(n-3)$ and $f_3 = n(n-3)/2$. The $\binom{n}{3}$ triples of vertices split into f_2 triangles and $\binom{n}{3} - f_2$ missing or empty triangles. Similarly the $\binom{n}{4}$ quadruples split into f_3 3-faces and several other types where in any case the span of this quadruple of vertices in M (= the corresponding 3-top-set) should have a first and second homology of rank at most 1. Hence there are left only two types:

1. an empty tetrahedron ($\cong S^2$),

2. the boundary of a tetrahedron with two faces removed ($\simeq S^1$).

Let A and B denote the number of quadruples of the first and second type, respectively. This leads to the following equations.

$$\begin{aligned} 4f_3 + 4A + 2B &= (n-3)f_2 \\ 2B &= (n-3)\left(\binom{n}{3} - f_2\right). \end{aligned}$$

Eliminating B, f_2, f_3 we get

$$24A = n(n-3)(n-5)(10-n).$$

By assumption A is nonnegative which implies $n \leq 10$. This means that we have to look for tight 2-neighborly triangulation with at most 10 vertices. By work of A. Altshuler and L. Steinberg [AS1], [Al2] there is a complete computer-aided classification of such 2-neighborly triangulations. The numbers of distinct combinatorial types are the following:

1	type for	$n \leq 7$	
4	types for	$n = 8$	
51	types for	$n = 9$	(50 spheres, 1 non-sphere)
3677	types for	$n = 10$	(3540 spheres, 137 non-spheres)

From 3.7 we know that the only tight triangulation of the 3-sphere has $n = 5$ vertices. The 137 non-spheres with 10 vertices are not tight because in each case there occur types of quadruples of vertices which are different from the types 1 and 2 above and which are therefore forbidden.

There remains exactly one candidate with 9 vertices. It is by construction the boundary of the nonorientable 1-handle defined as the union of the nine 4-simplices generated by

$$\langle 1\,2\,3\,4\,5\rangle$$

under the action of the cyclic shift $S = (123456789)$. In addition it is invariant under the reflection $R = (18)(27)(36)(45)$. Therefore the automorphism group is the dihedral group $D_9 = \langle S, R\rangle$. The set of the 27 tetrahedra of this 3-manifold M split into the two D_9-orbits of $\langle 1245\rangle$ and $\langle 1235\rangle$.

It remains to show that this triangulation is tight. It is sufficient to show that the span of each triple and quadruple of vertices (= each 2-top-set and 3-top-set of the canonical embedding) injects into M on the level of homology with coefficients in $\mathbf{Z}_2$. The boundary of each missing triangle is homologous to the generator of $H_1(M; \mathbf{Z}_2) \cong \mathbf{Z}_2$. There are 27 tetrahedra, 9 empty tetrahedra (in agreement with the formula $24A = n(n-3)(n-5)(10-n) = 9 \cdot 24$) whose boundaries represent the generator of $H_2(M; \mathbf{Z}_2) \cong \mathbf{Z}_2$. The 90 other quadruples are of type 2, and their homology injects also. The tightness of this example is also a special case of the more general Theorem 5.5 below.

5.4 Corollary: *There exists a tight and substantial polyhedral embedding of the 3-dimensional Klein bottle into E^8.*

In order to illustrate this triangulation and its canonical embedding into Δ^8, we consider a simplexwise linear function f on Δ^8 with $f(1) = f(2) = f(3) = f(4) = 1$, $f(5) = f(6) = f(7) = f(8) = -1$ and $f(9) = 0$. The combinatorial automorphism $R = (18)(27)(36)(45)$ preserves the 0-level and interchanges the $(+1)$-level and the (-1)-level. Figure 9 shows these three levels where $\frac{1}{5}$ denotes the centre of the edge joining 1 and 5, etc. The fixed point set of R is indicated by fat drawing. The topology of this manifold can be recognized from the two different identifications of the upper and the lower cylinder $S^2 \times [0, 1]$ interchanged by R.

level 1: boundary of a tetrahedron

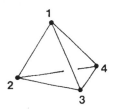

level 0

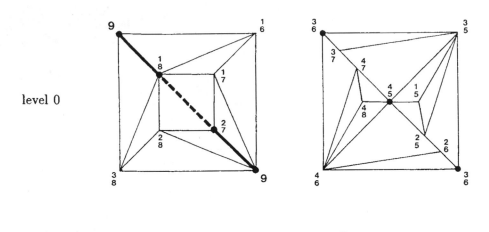

level -1: boundary of a tetrahedron

Figure 9

5B. Tight sphere products and twisted sphere bundles

As a generalization of 5.3, we show the following result on tight triangulations of sphere bundles over S^1:

5.5 Theorem: *For each $d \geq 2$ there exists a tight $(2d+3)$-vertex triangulation of a $(d+1)$-dimensional 1-handle which is orientable (or nonorientable) if d is even (or odd, respectively). Its boundary is a tight $(2d+3)$-vertex triangulation of the total space of the trivial (or twisted) S^{d-1}-bundle over S^1.*

PROOF: We define $n := 2d+3$ and regard the vertices as the elements of $\mathbf{Z}_n$. We define a simplicial complex N^{d+1} as the union of all elements in the $\mathbf{Z}_n$-orbit of the simplex spanned by

$$\langle 1\ 2\ 3\ \ldots\ d+2 \rangle.$$

This is a $(d+1)$-dimensional manifold with boundary. By construction it is a stacked $(d+1)$-polytope with identifications along two disjoint facets. Therefore it is PL homeomorphic to a 1-handle which is orientable if d is even and nonorientable if d is odd. Its boundary $M^d := \partial N^{d+1}$ is in Walkup's class $\mathcal{H}^{d+1}(1)$, see [Wa]. It is PL homeomorphic to $S^1 \times S^{d-1}$ if d is even and to the total space of the twisted S^{d-1}-bundle over S^1 ($= d$-dimensional Klein bottle) if d is odd. Compare [Ste] for the classification of sphere bundles. The link of each vertex of M^d is a stacked sphere, compare [Kal1;Sect.8], [Kü3].

It remains to show that M^d and N^{d+1} are tight triangulations. First of all, each of them is 2-neighborly and therefore 0-tight. N^{d+1} has the homotopy type of a circle. Applying 3.17 (iii) we will show that for any choice of a subset $A \subset \mathbf{Z}_n$ the span of A in N^{d+1} is either contractible or homotopy equivalent to N^{d+1} itself. Note that the generator of $\pi_1(N^{d+1}) \cong H_1(N^{d+1}; \mathbf{Z}) \cong \mathbf{Z}$ may be chosen as the union of all edges $\langle i\ i+1 \rangle$, $i \in \mathbf{Z}_n$.

Case 1: A is contained in a $(d+2)$-tuple of subsequent vertices of $\mathbf{Z}_n$. Then A spans a simplex of N^{d+1} which clearly is contractible.

Case 2: A is not contained in any $(d+2)$-tuple of subsequent vertices. Then its span collapses onto the union of three edges $\langle v_1 v_2 \rangle$, $\langle v_2 v_3 \rangle$, $\langle v_3 v_1 \rangle$ where any two of the three vertices v_1, v_2, v_3 (but not all) lie in a common $(d+2)$-tuple of subsequent vertices. It easily follows that the union of these three edges is homotopy equivalent to a generator of $\pi_1(N^{d+1})$.

Therefore that N^{d+1} is a tight triangulation. Similar considerations are possible for its boundary M^d. On the other hand it follows from 6.5 below that M^d is a tight triangulation since the relation $\sum_i b_i(M^d; \mathbf{Z}_2) = 4 = 2 \cdot \sum_i b_i(N^{d+1}; \mathbf{Z}_2)$ holds for $d \geq 2$. This relation fails to hold for $d = 1$ where N^2 is the tight 5-vertex Möbius band whose boundary is not tight.

REMARK: By construction N^{d+1} is a proper subcomplex of the cyclic polytope $C(2d+3, d+2)$ if d is even and of $C(2d+3, d+3)$ if d is odd. In particular this implies that for even d M^d can be polyhedrally embedded as a hypersurface of E^{d+1} (take the Schlegel-diagram of the polytope). For $d = 2$ this is known as Császár's torus [Cs], [Al1]; the other examples are higher dimensional analogues, see [Kü3]. Observe that for even d the closure of the complement of N^{d+1} in $\partial C(2d+3, d+2)$ is again a combinatorial manifold with the same boundary. However, it is not a tight triangulation for $d \geq 4$ although the boundary is tight and the relation for the homology is satisfied, compare 6.5.

5.6 Corollary: *For each $d \geq 2$ there is a tight and substantial polyhedral embedding*

$$S^1 \times S^{d-1} \to E^{2d+2} \text{ if } d \text{ is even}$$

and of the twisted bundle

$$S^1 \underline{\times} S^{d-1} \to E^{2d+2} \text{ if } d \text{ is odd}.$$

Compare the tight and taut algebraic embedding of the same manifold into E^{2d} which is given by a focal set of a certain isoparametric hypersurface in S^{2d-1} with four distinct principal curvatures [CR;Ex.7.4]. This example can be described as the complexified sphere S^{d-1} in $\mathbf{C}^d$ or, alternatively, as a tensor product $S^1 \otimes S^{d-1}$ [Kü9]:

$$f(t, x) := e^{it} \cdot x \text{ for } x \in S^{d-1}.$$

Compare also 6.9 below for tight embeddings of the odd-dimensional products and the even-dimensional twisted products.

5.7 Proposition ([Br-Kü3]): *The examples M^d of Theorem 5.5 have the minimum number of vertices among all d-manifolds which are not simply connected, $d \geq 3$.*

SKETCH OF PROOF: Let n denote the number of vertices. One single top-dimensional simplex requires already $d + 1$ vertices. The complex spanned by the other $n - d - 1$ vertices must contain all the critical points of index $1, ..., d-1$ for any simplexwise linear function. If the manifold is not simply connected then there must be at least one critical point of index 1 and, by duality, at least one critical point of index $d - 1$ whenever $d \geq 3$. The latter one requires at least $d + 1$ vertices, and the presence of the critical point of index 1 requires one extra vertex. Altogether we get $n - d - 1 \geq d + 2$.

6. Connected sums and manifolds with boundary

Let us first recall T. Banchoff's construction [Ba5] of a tight polyhedral Klein bottle in E^5. We start with the tight 6-vertex triangulation of $\mathbf{R}P^2$ and remove the open star of one vertex. This leads to the tight 5-vertex triangulation of the Möbius band (Figure 2). Then we take two translational copies of the canonical embedding of this 5-vertex Möbius band in two parallel hyperplanes of E^5. Finally we join the boundaries by a straight cylinder. This embedding of $\mathbf{R}P^2 \sharp \mathbf{R}P^2$ (= Klein bottle) into E^5 is tight by 2.10 since it is 1-Hamiltonian in the boundary of the prism $\triangle^4 \times [0,1]$. In the following chapter we generalize this construction in several ways. This generalized construction leads, among others, to tight polyhedral embeddings of connected sums of copies of $\mathbf{C}P^2$ into E^8.

6A. Manifolds with one hole and connected sums

In this section we investigate partial analogues of the results of Section 2E for 2-dimensional manifolds with boundary. One of the basic facts says that tightness in presence of a nonempty boundary implies that all the extreme vertices lie on the boundary.

6.1 Proposition: *Let $M \subset E^d$ be a tight polyhedral manifold with boundary. Then ∂M contains all vertices of $\mathcal{H}(M)$, thus $\mathcal{H}(\partial M) = \mathcal{H}(M)$. If M is a tight triangulation then each vertex of M is in ∂M.*

PROOF: Assume that v is a vertex of $\mathcal{H}$ with $v \in M \setminus \partial M$. There is a half space h containing all vertices of $\mathcal{H}$ except v. It follows that $M \setminus h$ is topologically a ball around v. Therefore $M \cap \partial h$ bounds in M. On the other hand $M \cap \partial h$ represents a non-vanishing homology class in $M \cap h$ which is a manifold with boundary and with an additional hole. Therefore

$$H_*(M \cap h) \to H_*(M)$$

ist not injective. This contradicts the tightness, independently of the field of coefficients. For a tight triangulation we just have to consider the canonical embedding where each vertex of M is also a vertex of $\mathcal{H}$. For the case of 2-manifolds compare 2.23.

6.2 Proposition: *Let M be a compact n-manifold without boundary, and define M' to be M with an open n-ball removed. Then every tight triangulation of M induces a tight triangulation of M' by cutting out the open star of one vertex. Vice versa, for even $n = 2k$ a tight triangulation of M' with respect to $\mathbf{Z}_2$ induces a tight triangulation of M by adding the cone over $\partial M'$ to an additional vertex, provided that $\partial M'$ is triangulated as a k-neighborly combinatorial sphere.*

PROOF: The canonical embedding of the triangulated M' appears as a top-set of the canonical embedding of the triangulated M and is therefore tight, see 1.4. Vice versa, let $n = 2k$ and assume that the canonical embedding of the triangulated M' into, say, $\triangle^d$ is tight, and consider the canonical embedding of the triangulated M into $\triangle^{d+1} = \triangle^d * \{v\}$. Let $h \subset E^{d+1}$ be a half space. If h does not contain the additional vertex v then $h \cap M$ retracts to $h \cap M'$, and

$$H_*(h \cap M) \cong H_*(h \cap M') \to H_*(M') \to H_*(M)$$

is injective. If $h \cap M' = \emptyset$ then $h \cap M$ is either empty or a ball around v which has trivial homology. There remains the case where $v \in h$ and $h \cap M'$ is a nonempty proper subset of M'. In this case $h \cap M$ is the union of $h \cap M'$ and the cone over $h \cap \partial M'$ with apex v. By the k-neighborliness of $\partial M'$ $h \cap \partial M'$ is $(k-2)$-connected. It follows that adding the cone over $h \cap \partial M'$ with apex v does not contribute to the homology $H_1, \ldots, H_{k-1}$. It follows that

$$H_i(h \cap M) \cong H_i(h \cap M') \to H_i(M') \to H_i(M)$$

is injective for $i = 0, 1, \ldots, k-1$. This shows that M is $(k-1)$-tight with respect to $\mathbf{Z}_2$. By 3.18 it is tight with respect to $\mathbf{Z}_2$.

EXAMPLE: There is a tight 8-vertex triangulation of the total space of the nontrivial 2-disc bundle over S^2 whose boundary is the Hopf bundle: Take $\mathbf{C}P_9^2$ minus an open vertex star according to 4.13.

6.3 Theorem: *If a $(k-1)$-connected $2k$-manifold M without boundary admits a tight triangulation with n vertices, then there is a tight and substantial polyhedral embedding*

$$M \sharp (-M) \to E^{n-1}$$

where $-M$ denotes the same manifold with the opposite orientation.

PROOF: By 6.2 there is a tight triangulation of M' with $n - 1$ vertices where M' is M minus an open ball. Let us take two translationally congruent copies of the canonical embedding of M' in two parallel hyperplanes of E^{n-1}. These can be joined by a straight cylinder $\partial M' \times [0, 1]$. By assumption the triangulation of M (and hence of M') is $(k + 1)$-neighborly, and the subcomplex $\partial M'$ is k-neighborly. This implies that $M \sharp (-M)$ is k-Hamiltonian in the prism and hence tight in E^{n-1} by 4.1.

6.4 Corollary: *There exists a tight and substantial polyhedral embedding*

$$\mathbf{C}P^2 \sharp (-\mathbf{C}P^2) \to E^8.$$

This follows from 6.3 and 4.13. By a theorem of G. Thorbergsson [Th1] there is no smooth tight immersion of $\mathbf{C}P^2 \sharp (-\mathbf{C}P^2)$ into any Euclidean space. Applying the construction of 6.3 to a tight triangulation of a surface of genus g (cf. 2.16) where $\binom{n-3}{2} = 6g$, we obtain a tight embedding of the surface of genus $2g$ into the prism $\triangle^{n-2} \times [0, 1]$.

6B. Manifolds with boundary

The relationship between the tightness of manifold and the tightness ot its boundary certainly depends on the homology of the space and its boundary. This section gives a few results in this direction.

6.5 Proposition: *Assume that $M \subset E^d$ is a tight polyhedral embedding of a manifold with boundary such that the F-Betti numbers satisfy*

$$\sum_i b_i(\partial M) = 2 \sum_i b_i(M).$$

Then $\partial M \subset E^d$ is also tight.

In particular, under the same assumption the boundary of a tight triangulation is also a tight triangulation.

PROOF: Let $z^* : M \to \mathbf{R}$ be a height function in general position, and let z_0^* and z_1^* denote its restriction to $M \setminus \partial M$ and ∂M, respectively. The following inequality was proved in [Kü1;3.3] as a consequence of the Alexander duality for the vertex links:

$$\mu_i(z^*) + \mu_i(-z^*) \geq \mu_i(z_0^*) + \mu_i(-z_0^*) + \mu_i(z_1^*)$$

which by the tightness and by the Morse inequality 3.15 implies

$$2b_i(M) = \mu_i(z^*) + \mu_i(-z^*) \geq \mu_i(z_1^*) \geq b_i(\partial M)$$

for each i. The assumption about the Betti numbers implies

$$\sum_i \mu_i(z_1^*) = \sum_i b_i(\partial M),$$

hence ∂M is tight.

6.6 Corollary: *The image of any tight polyhedral embedding of an n-ball into Euclidean space is a convex n-polytope in E^n, and any tight triangulation of an n-ball is one n-simplex with $n+1$ vertices.*

This follows from 6.5, 3.6 and 3.7 because of $\sum_i b_i(S^{n-1}) = 2 = 2\sum_i b_i(B^n)$, where B^n denotes an n-ball. In fact, 6.6 holds also for homology d-balls.

REMARK: The converse of 6.5 is not true in general. The tightness of the triangulation ∂M does not imply the tightness of the triangulation M, not even if all vertices lie on ∂M and if the relation for the homology is satisfied, compare the example in the remark after 5.5. However, this converse is true for 3-manifolds with connected boundary, according to the following theorem:

6.7 Theorem: *Let M be a combinatorial 3-manifold with nonempty and connected boundary, and assume that ∂M is a tight triangulation and that ∂M contains all vertices of M. Then M is also a tight triangulation and moreover $\sum_i b_i(\partial M) = 2\sum_i b_i(M)$ and $b_2(M) = 0$ where we have to use coefficients in $\mathbf{Z}_2$ if ∂M is nonorientable. Furthermore there is a tight and substantial embedding*

$$M \cup_{\partial M} M \to E^n$$

as a subcomplex of the prism $\triangle^{n-1} \times [0,1]$, where n denotes the number of vertices of M.

PROOF: By assumption ∂M is a tight triangulation, and 2.16 implies that it is 2-neighborly. This means that $M \setminus \partial M$ does not contain vertices or edges. In particular the link of each vertex is a triangulated 2-disc (an $(n-1)$-gon) with no vertices in the interior. Now let z^* be a height function in general position on the canonical embedding $M \to \triangle^{n-1}$, and let z_1^* denote its restriction to ∂M. Then we claim that for each i

$$\mu_i(z_1^*) = \mu_i(-z_1^*) = \mu_i(z^*) + \mu_i(-z^*).$$

There is nothing to prove for $i \geq 3$. For $i = 0$ the equation is trivial because a maximum lying on ∂M is not critical for the function on M. Similarly, for the case $i = 2$ we just have to interchange the roles of minima and maxima. For $i = 1$ the equation follows by counting the number of components in a vertex link L (which is a 2-disc) and in $\partial L = L \cap \partial M$) which lie above and below the level of the corresponding vertex. The contribution to $\mu_1(z^*)$ or $\mu_1(z_1^*)$ is just the number of $(-)$-components minus 1, and $\mu_1(-z^*)$ or $\mu_1(-z_1^*)$ is the number of $(+)$-components minus 1.

For 3-manifolds with boundary in general we have

$$2\chi(M) = \chi(\partial M)$$

and hence

$$
\begin{aligned}
2\sum_i b_i(M) &= 2 + 2b_1(M) + 2b_2(M) \\
&\geq 2 + 2b_1(M) - 2b_2(M) \\
&= 4 - 2\chi(M) \\
&= 4 - \chi(\partial M) \\
&= 2 + b_1(\partial M) \\
&= \sum_i b_i(\partial M)
\end{aligned}
$$

with equality if and only if $b_2(M) = 0$.

This implies

$$
\begin{aligned}
\sum_i b_i(\partial M) = \sum_i \mu_i(z_1^*) &= \sum_i(\mu_i(z^*) + \mu_i(-z^*)) \\
&\geq 2\sum_i b_i(M) \\
&\geq \sum_i b_i(\partial M).
\end{aligned}
$$

Consequently we have equality at each step which implies that M is tight and $b_2(M) = 0$ and $b_1(\partial M) = 2b_1(M)$. If ∂M is an orientable surface then M is a classical handlebody of the same genus as ∂M.

The construction of the embedding of $2M := M \cup_{\partial M} M$ into E^n is essentially the same as in 6.3 above: take two translationally congruent copies of the canonical embedding $M \subset \triangle^{n-1}$ in two parallel hyperplanes of E^n and join them by a cylinder $\partial M \times [0, 1]$. To see the tightness we first observe that

$$H_1(2M) \cong H_1(M)$$

because the two copies are glued together via the identity on ∂M. Secondly, let h be a half space in E^n. By construction $h \cap 2M$ is connected, therefore the embedding in 0-tight. Furthermore $h \cap \partial M$ is connected and hence $h \cap (\partial M \times [0, 1])$ does not contribute elements to $H_1(h \cap 2M)$ which are not already represented in one of the two copies of $h \cap M$. But these inject into $H_1(M)$ by the tightness of M. It follows that $2M \to E^n$ is 1-tight and hence tight by 3.18. If ∂M is orientable then this holds for any field F.

6.8 Examples:

1. The 7-vertex triangulation of a solid torus is tight by 6.7 because its boundary is tight. It is also tight by 5.5. The last part of 6.7 leads to a tight polyhedral embedding $S^1 \times S^2 \to E^7$ as a subcomplex of the prism $\triangle^6 \times [0, 1]$. No example seems to be known of a tight and substantial embedding of $S^1 \times S^2$ into E^8. Compare, however, the tight 3-dimensional Klein bottle in E^8 (see 5.4). Compare also 6.9.

2. There is the following tight 19-vertex triangulation of the orientable surface of genus $g = 20$ (see [Ri3;2.3]). Regard the vertices as elements of $\mathbf{Z}_{19}$ and take the union of the $\mathbf{Z}_{19}$-orbits of the following triangles:

$$\langle 016 \rangle, \quad \langle 026 \rangle, \quad \langle 038 \rangle, \quad \langle 047 \rangle, \quad \langle 079 \rangle, \quad \langle 089 \rangle.$$

This is the boundary of a 3-manifold M_{20} defined as the union of the $\mathbf{Z}_{19}$-orbits of the tetrahedra

$$\langle 0126 \rangle, \quad \langle 0348 \rangle, \quad \langle 0478 \rangle, \quad \langle 0789 \rangle.$$

M_{20} is a tight triangulation by 6.7. Topologically it is an orientable handlebody of genus $g = 20$, and the last part of 6.7 leads to a tight polyhedral embedding of the connected sum of 20 copies of $S^1 \times S^2$ into E^{19}.

3. (U. Brehm, personal communication):
   Consider 12 vertices $1, 2, 3, 4, A, B, C, D, a, b, c, d$ and the C_4-action defined by $(1234)(ABCD)(abcd)$. The union N of the orbits of the tetrahedra

$$\langle ABCD \rangle, \quad \langle ABC3 \rangle, \quad \langle aBC3 \rangle, \quad \langle adC3 \rangle,$$

$$\langle adD3 \rangle, \quad \langle 2dC3 \rangle, \quad \langle 2bd3 \rangle, \quad \langle 3ac1 \rangle,$$

is a tight triangulation of a 3-manifold whose boundary is a tight 12-vertex triangulation of the orientable surface of genus 6. It is an orientable handlebody of genus 6, and 6.7 induces a tight polyhedral embedding of the connected sum of six copies of $S^1 \times S^2$ into E^{12}. A second, essentially different, copy of such a handlebody is the union N' of the orbits of the tetrahedra

$$\langle abcd \rangle, \quad \langle abcA \rangle, \quad \langle a4cA \rangle, \quad \langle D4cA \rangle,$$

$$\langle D4c3 \rangle, \quad \langle D42A \rangle, \quad \langle 2BD4 \rangle, \quad \langle 3AC1 \rangle,$$

$\partial N = \partial N'$ is the same tight triangulation of the surface of genus 6, and $N \cup N'$ is a triangulation of S^3. This follows from the decomposition of the set of vertices into $\{A, B, C, D, 1, 3\}$ and $\{a, b, c, d, 2, 4\}$, each spanning a collapsible subcomplex.

A suitable convex embedding of this triangulated S^3 into E^4 would solve the long-standing problem whether or not a 12-vertex triangulation of an orientable surface of genus 6 can be simplexwise linearly embedded into E^3, compare [Br1], [Bo Br]. None of the 12-vertex triangulations of this surface can be a subcomplex of a convex 4-polytope according to [ABS2]. It is not true that every triangulated surface occurs as the boundary of a triangulated 3-manifold without additional vertices, see [Br2].

6.9 Proposition: *For each $d \geq 3$ there is a tight and substantial polyhedral embedding*

$$S^1 \times S^{d-1} \to E^{2d+1} \quad \textit{if d is odd}$$

and of the twisted bundle

$$S^1 \underline{\times} S^{d-1} \to E^{2d+1} \quad \textit{if d is even.}$$

In any case the manifold may be chosen as a subcomplex of a prism $\triangle^{2d} \times [0, 1]$.

PROOF: We use the same construction as in 6.3 or 6.7. By 6.5 there is a tight $(2d+1)$-vertex triangulation of $S^1 \times B^{d-1}$ if d is odd and of $S^1 \underline{\times} B^{d-1}$ if d is even. We take two copies of this tight subcomplex of $\triangle^{2d}$, regarded as translational copies in parallel hyperplanes of E^{2d+1}. Then we join the two boundaries by a straight cylinder. This yields a tight embedding

$$S^1 \times B_+^{d-1} \cup S^1 \times B_-^{d-1} \cong S^1 \times S^{d-1} \longrightarrow \triangle^{2d} \times [0,1]$$

or

$$S^1 \underline{\times} B_+^{d-1} \cup S^1 \underline{\times} B_-^{d-1} \cong S^1 \underline{\times} S^{d-1} \longrightarrow \triangle^{2d} \times [0,1],$$

respectively. The tightness follows by the same argument as in the proof of 6.7, using the particularly simple homology of this space.

6C. Connected sums by iterated truncation

In order to get tight embeddings of connected sums of more than two copies of a given manifold we have to modify the concept used in 6.3 and 6.7. Let us go back to the 2-dimensional case. We start again with the canonical embedding of the 6-vertex $\mathbf{R}P^2$ into $\triangle^5$ and truncate $\triangle^5$ at one vertex. This gives an embedded $\mathbf{R}P^2$ minus a small pentagon where the five vertices of the pentagon lie in general position. This enables us to attach a small 5-vertex Möbius band to the boundary. The result is a polyhedral embedding of the Klein bottle into E^5 which is 1-Hamiltonian in the truncated simplex and hence tight (in fact, it is projectively equivalent to Banchoff's tight Klein bottle in the prism $\triangle^4 \times [0,1]$).

However, this process of truncation can be repeated arbitrarily often since each time the boundary after truncation is a pentagon. If we start with a regular 5-simplex, truncate each of the six vertices in an appropriate way, and attach six Möbius bands as described above we get a tight polyhedral embedding

$$\mathbf{R}P^2 \sharp 6\mathbf{R}P^2 \to E^5$$

whose image is invariant under the full automorphism group ($\cong A_5$) of the 6-vertex $\mathbf{R}P^2$, see Figure 10. This is 1-Hamiltonian in the truncated 5-simplex all of whose edges have the same length, a kind of Archimedean tessellation with triangles and hexagons. Anyhow, the same construction can be carried out by starting with the canonical embedding of any tight triangulation of a surface. Our next goal is to show that this is possible also in higher dimensions.

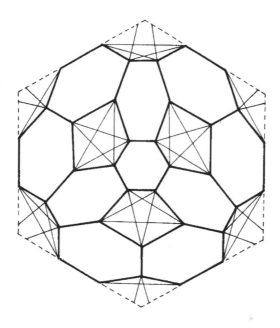

Figure 10

6.10 Theorem: *Let M be a tight triangulation of a $(k-1)$-connected $2k$-manifold with n vertices. Then for an arbitrary integer $m \geq 0$ there is a tight and substantial polyhedral embedding*

$$M \sharp m(-M) \to E^{n-1}.$$

PROOF: We start with the canonical embedding $M \to \triangle^{n-1}$. Truncating one vertex v of $\triangle^{n-1}$ leads to an embedding of M with one hole where the boundary is a subcomplex of a small $(n-2)$-simplex in the slicing hyperplane. If we take a second copy of the triangulation of M minus the open star of v, we find that the two boundaries are combinatorially equivalent where a vertex w of $M \setminus$ star (v) corresponds to the intersection point w' of the edge vw with the slicing hyperplane. This enables us to put in $M \setminus$ star (v) via the canonical embedding into this small $(n-2)$-simplex, identifying w with w': the result is a polyhedral $M \sharp (-M)$ embedded as a subcomplex of the (once) truncated $\triangle^{n-1}$ which is k-Hamiltonian and hence tight by 4.1. This procedure can be repeated n times.

In order to go further we have to look what happens if we truncate at one of the new vertices w' arising from the intersection with the original edges. First of all the truncated simplex is a simple polytope meaning that the vertex figure at each vertex is a simplex (Def. 1.1). The vertex figure of w' consists of simplices of two kinds, namely the old ones from M and the new ones from the small copy of $M \setminus \mathrm{star}\ (v)$. The set of the old ones is combinatorially isomorphic to

$$w * \mathrm{link}(vw),$$

the set of the new ones is combinatorially isomorphic to

$$\mathrm{link}(w) \setminus \mathrm{star}\ (vw),$$

thus the union of both parts is combinatorially isomorphic to the link of w in M.

This makes is possible to truncate at the new vertices of type w' and to put in a copy of $M \setminus \mathrm{star}\ (w)$. The same procedure can be carried out also for the vertices arising from this truncation of the second level. It follows that the number of steps is not bounded. At each step the resulting manifold is k-Hamiltonian in the repeatedly truncated simplex Δ^{n-1} and is therefore tight by 4.1.

6.11 Corollary: *For an arbitrary integer $m \geq 0$ there is a tight polyhedral embedding $\mathbf{C}P^2 \sharp m(-\mathbf{C}P^2) \to E^8$.*

There are no smooth tight immersions of this these manifolds into E^8 except for $m = 0$. For $m \leq 2$ there are no smooth tight immersions into E^6; for $m \geq 3$ smooth tight immersions into E^6 or E^7 are not known but might be possible for large m, see [Th1].

6.12 Corollary: *For an arbitrary integer $m \geq 0$ there is a tight polyhedral embedding $M^8 \sharp m(-M^8) \to E^{14}$ where M^8 is the 8-manifold 'like the quaternionic projective plane' from 4.18.*

7. Miscellaneous cases and pseudomanifolds

In Section 7A we extend the results of Section 5A to some higher dimensional cases by essentially the same methods. We obtain a characteristic property of the tight 11-vertex triangulation of $S^1 \times S^3$. In the second part we consider $(k-1)$-connected $(2k-1)$-dimensional pseudomanifolds. In a certain sense these may be regarded as analogues of the $(k-1)$-connected $2k$-manifolds — combinatorially as well as topologically. A number of results of Chapter 4 can be carried over to this case. In particular, we obtain an inequality of the Heawood type, where again the case of equality coincides with the case of tight triangulations. We come back to polyhedral manifolds at the very end: Slices by hyperplanes in general position can lead to tight polyhedral $2k$-manifolds which are $(k-1)$-connected. We call these tight slicings.

7A. Special cases in dimensions 4 and 5

7.1 Proposition: *Let M be a tight n-vertex triangulation of a 4-manifold with rk $H_1(M) \leq 1$. Then the following holds*

$$n(n-11)(n-13) \leq 120\chi(M)$$

with equality for $n \geq 7$ if and only if no 4-tuple of vertices spans the boundary of a tetrahedron.

PROOF: The proof follows the pattern of 5.3. The assumption implies $\chi(M) \geq b_2(M) \geq 0$. Let A denote the number of 4-tuples spanning the boundary of a tetrahedron ($=$ an empty 3-simplex, 5.1). By the same method as in 5.3 one calculates

$$0 \leq 24A = (n-6)[120\chi(M) - n(n-11)(n-13)]$$

which proves the assertion for $n \geq 7$. For $n = 6$ M is the unique tight triangulation of S^4; thus the inequality is still true.

7.2 Proposition (D. Walkup, [Wa;Thm.5]): *For any combinatorial 4-manifold* M *with* n *vertices the inequality*

$$\binom{n-5}{2} \geq \frac{15}{2}(2 - \chi(M))$$

holds with equality if and only if the triangulation is 0-tight and each vertex link is a stacked sphere (5.1).

PROOF: For the link L_i of the vertex v_i we have the Lower Bound Theorem

$$f_1^{(i)} \geq 4f_0^{(i)} - 10.$$

Summation over i leads to

$$3f_2 \geq 8f_1 - 10n.$$

Combining this with the equation $f_2 - 4f_1 + 10n = 10\chi(M)$ (see Section 4B), we get

$$
\begin{aligned}
30\chi(M) &= 3f_2 - 12f_1 + 30n \\
&\geq 8f_1 - 10n - 12f_1 + 30n \\
&= 20n - 4f_1 \\
&\geq 20n - 4\binom{n}{2} \\
&= -4\binom{n-5}{2} + 60
\end{aligned}
$$

with equality if and only if $f_1 = \binom{n}{2}$ and $f_1^{(i)} = 4f_0^{(i)} - 10$ for each i. Hence the triangulation is 0-tight, and each vertex link satisfies equality in the Lower Bound Theorem, see 5.1.

REMARK: Equality in 7.1 and 7.2 is attained for the tight 11-vertex triangulation of $S^1 \times S^3$, see 5.5. If we assume $H_2(M) = 0$ then Walkup's inequality becomes

$$\binom{n-5}{2} \geq 15 \cdot b_1(M),$$

and it states $n \geq 11$ for $b_1(M) = 1$. Compare also 5.7 and the bounds given in [Br-Kü3].

A tight triangulation of $\mathbf{R}P^4$ does not exist: By 7.1 it would have at most 15 vertices, and by the condition about empty 3-simplices it would have exactly 15 vertices. This follows because it is impossible that an empty 3-simplex represents the generator of $H_2(\mathbf{R}P^4; \mathbf{Z}_2)$. In fact, this generator would have to be a real projective plane (up to homology) and therefore would require at least 6 vertices. Generalizing this idea, an induction given in [AM] shows that any triangulation of $\mathbf{R}P^d$ would require $n \geq \binom{d+2}{2}$ vertices and that this bound is not attained for $d \geq 3$ since there is no 10-vertex triangulation of $\mathbf{R}P^3$ (compare 5.3).

7.3 Proposition: *Let M be a tight n-vertex triangulation of a simply connected 5-manifold with $rk\ H_2(M) \leq 1$. Then $n \leq 13$ holds with equality if and only if no 5-tuple of vertices spans the boundary of a 4-simplex.*

PROOF: As is 7.1, the number A of empty 4-simplices satisfies the relation

$$0 \leq 80A = n(n-4)(n-6)(n - 7)(13-n).$$

REMARK: A tight triangulation of $S^2 \times S^3$ would have at most 12 vertices because the topology seems to require missing 4-simplices. On the other hand any triangulation of $S^2 \times S^3$ must have at least 12 vertices [Br-Kü3].
Similar considerations for simply connected 6-manifolds M with $rk\ H_2(M) \leq 1$ show that a tight triangulation of $S^2 \times S^4$ or of CP^3 would have either 14 or 16 vertices. According to [AM] CP^d would require $n \geq (d+1)^2$ vertices but this is not attained for $d \geq 3$. By [Br-Kü3] any triangulation of $S^2 \times S^4$ must have at least 14 vertices.

7B. Triangulated pseudomanifolds

7.4 Definition: A combinatorial n-pseudomanifold with isolated singularities is a simplicial complex such that the link of each vertex is a connected combinatorial $(n-1)$-manifold. A polyhedral d-pseudomanifold with isolated singularities in E^d is a polyhedron such that the link of each vertex is a connected polyhedral $(n-1)$-submanifold. A vertex is called a proper singularity if its link is not a sphere. The connectedness of each vertex link implies that the complex itself is strongly connected, i.e., any two top-dimensional simplices can be joined by a chain of such, compare Section 2F.

Explicit examples of such triangulated pseudomanifolds can be found in [Al4], [Kü4] or in 7.16 below. 7.2 holds by the same proof also for 4-dimensional pseudomanifolds.

7.5 Examples (Tightly embedded pseudomanifolds): *Let K be any combinatorial $(n-1)$-manifold with d vertices. Then $2^K \subset C^d \subset E^d$ is a tightly embedded n-pseudomanifold with isolated singularities , each of its vertices being a proper singularity if K is not a sphere.*

If $K = \{3,5\}_5$ is the 6-vertex $\mathbb{R}P^2$ or if $K = \{3,6\}_{1,2}$ is the 7-vertex torus then 2^K is an abstract regular 4-polytope of Schläfli type $\{4,3,5\}$ or $\{4,3,6\}$, respectively, compare [MM-S]. This 3-pseudomanifold in C^6 (or C^7, respectively) is 2-Hamiltonian and therefore simply connected.

In general, if K is k-neighborly and $n = 2k - 2$ then 2^K is a k-Hamiltonian pseudomanifold of C^d which is $(k-1)$-connected by 3.8. For the particular case of $K = \mathbb{C}P_9^2$ (see 4.13) $2^K \subset E^9$ is a tight 2-connected 5-pseudomanifold with 512 vertices, each being of the type 'cone over $\mathbb{C}P^2$'. Tight slicings of this pseudomanifold can produce simply connected 4-manifolds, tightly embedded by 7.17.

In the simplicial case a combinatorial $(2k-1)$-pseudomanifold with isolated singularities can be $(k+1)$-neighborly (which is impossible for $(2k-1)$-manifolds except for the trivial case of $\partial\Delta^{2k}$). Therefore in a combinatorial sense these are odd-dimensional analogues of the tight triangulations of $(k-1)$-connected $2k$-manifolds considered in Chapter 4. For 3-dimensional examples of tight triangulations see 7.16 below. 7.2 holds also for pseudomanifolds with isolated singularities.

7.6 Proposition (Duality for tightness:) *Assume there is a polyhedral embedding of a $(k-1)$-connected $(2k-1)$-pseudomanifold M with isolated singularities in E^d such that the link of each vertex is a $k-2$-connected $(2k-2)$-manifold. Then the following conditions are equivalent:*

(i) The embedding is tight.

(ii) The embedding is $(k-1)$-tight.

PROOF: We may assume that $k \geq 2$. (i) $\Rightarrow$ (ii) is trivial. In order to show (ii) $\Rightarrow$ (i) we use the same method as in the proof of 3.18. The main problem is that the Poincaré duality does not hold for pseudomanifolds and that the Alexander duality does not hold for the vertex links, at least not in the classical form which we would need for the validity of 3.15 (vi) in this case. By assumption for any height function in general position we have $\mu_0(z) = 1, \mu_1(z) = \ldots = \mu_{k-1}(z) = 0$. In view of the Euler-Poincaré equation 3.15 (iv)

$$\sum_i (-1)^i \mu_i = \sum_i (-1)^i b_i = \chi(M)$$

it is sufficient to show $\mu_{k+1}(z) = \ldots = \mu_{2k-2}(z) = 0, \mu_{2k-1}(z) = 1$.

First of all, we have $\mu_{2k-1} = 1$ because a vertex v is critical of index $2k-1$ if and only if it is the absolute maximum. The strong connectedness of M implies that the maximum has multiplicity 1.

Secondly, a vertex v is noncritical of index $j \geq 2$ if and only if $H_j(M_v, M_v \setminus v) = 0$ or, equivalently, $H_{j-1}(M_v \cap \mathrm{link}(v)) = 0$.

By the Poincaré-Lefschetz duality 3.14 (iii) for $\mathrm{link}(v)$ we get

$$H_{j-1}(M_v \cap \mathrm{link}(v)) \cong H_{2k-2-j+1}(\mathrm{link}(v), \mathrm{link}(v) \setminus M_v).$$

By assumption we have $H_1(\mathrm{link}(v)) = \ldots = H_{k-2}(\mathrm{link}(v)) = 0$. From the long exact sequence

$$\ldots \longrightarrow H_i(\mathrm{link}(v)) \longrightarrow H_i(\mathrm{link}(v), \mathrm{link}(v) \setminus M_v) \longrightarrow$$
$$\longrightarrow H_{i-1}(\mathrm{link}(v) \setminus M_v) \longrightarrow H_{i-1}(\mathrm{link}(v)) \longrightarrow \ldots$$

we obtain $\quad H_{2k-2-j+1}(\mathrm{link}(v), (\mathrm{link}(v) \setminus M_v) \cong H_{2k-2-j}(\mathrm{link}(v) \setminus M_v) = 0$

for $1 \leq 2k - 2 - j \leq k - 3$. Here the assumption about $(k-1)$-tightness comes in saying that $\mu_1(-z) = \ldots = \mu_{k-1}(-z) = 0$; we use this as a substitute for the duality 3.15 (vi).

Combining these arguments, we see that no vertex v is critical of index j for $1 \leq 2k - 2 - j \leq k - 3$, equivalently $k + 1 \leq j \leq 2k - 3$. For the case $j = 2k - 2$ we observe that in the long exact sequence above $H_0(\mathrm{link}(v) \setminus M_v) \to H_0(\mathrm{link}(v))$ is an isomorphism because $\mathrm{link}(v) \setminus M_v$ is connected.

Hence $H_1(\mathrm{link}(v), \mathrm{link}(v) \setminus M_v) = 0$ which implies $\mu_{2k-2}(z) = 0$. Altogether we obtain $\mu_{k+1}(z) = \ldots = \mu_{2k-2}(z) = 0, \mu_{2k-1}(z) = 1$. As a by-product, this implies $H_{k+1}(M) = \ldots = H_{2k-2}(M) = 0$, an analogue of the Poincaré duality for M. All considerations hold for any field F as coefficients, except possibly if $k = 2$. In this case we can use $F = \mathbf{Z}_2$.

7.7 Corollary: *A triangulation of a $(k-1)$-connected $(2k-1)$-pseudomanifold with isolated singularities is tight if and only if it is $(k-1)$-tight.*

PROOF: By 3.10 the triangulation is $(k-1)$-tight if and only if it is $(k+1)$-neighborly. This implies that each vertex link is k-neighborly and thus $(k-2)$-connected. If we regard the canonical embedding of the pseudomanifold, then 7.6 implies that the triangulation if tight.

7.8 Proposition (Dehn-Sommerville equations): *Let M be a combinatorial $(2k-1)$-pseudomanifold with isolated singularities . With the same notations as in 4.3 the Dehn-Sommerville equations hold as follows:*

$$\sum_{i=0}^{2k-1} (-1)^i f_i = \chi(M)$$

$$\sum_{i=2j}^{2k-1} (-1)^i \binom{i+1}{2j} f_i = 0 \quad for \ 1 \le j \le k-1$$

or, equivalently,

$$h_j - h_{2k-j} = (-1)^j \left(\binom{2k-1}{j} - \binom{2k-1}{j-1} \right) \chi(M) \quad for \ 0 \le j \le k-1.$$

The proof is essentially the same as in 4.3. Note that all odd-dimensional links are spheres. The only difference is that the Euler characteristic of an odd-dimensional pseudomanifold may be different from zero.

7C. Generalized Heawood inequalities for pseudomanifolds

In this section we present analogues of the generalized Heawood inequalities for $(k-1)$-connected $2k$-manifolds in 4.5, 4.6 and 4.7. In particular we obtain a characterization of tight triangulations of $(k-1)$-connected $(2k-1)$-pseudomanifolds.

7.9 Theorem: *Let P be a simplicial d-polytope and let $M \subset P \subset E^d$ be a tight subcomplex which contains all vertices of P and which is a $(k-1)$-connected combinatorial $(2k-1)$-pseudomanifold with isolated singularities . Then the following holds:*

$$(d+1) \cdot \binom{d-k-1}{k} \le (-1)^k \binom{2k}{k} \chi(M) = \binom{2k}{k} b_k(M).$$

Moreover, for $d \ge 2k+1$ equality holds if and only if P is a simplex (and, consequently, if M is a tight triangulation).

Note that the assumptions of 7.9 imply that each vertex link is $(k-1)$-Hamiltonian in the vertex figure; therefore it is a $(k-2)$-connected $(2k-2)$-manifold. By the proof of 7.6 this implies that $b_i(M) = 0$ for any $i \ne 0, k, 2k-1$. Therefore the right hand side of the inequality above is nonnegative: $(-1)^k \chi(M) = b_k(M)$.

7.10 Theorem: *Assume that M is a $(k-2)$-tight triangulation of a $(k-2)$-connected $(2k-1)$-pseudomanifold with isolated singularities and with n vertices. Then the following holds:*

$$n \cdot \binom{n-k-2}{k} \geq (-1)^k \binom{2k}{k} \chi(M)$$

with equality if and only if M is $(k-1)$-connected and the triangulation is tight.

7.11 Corollary: *An n-vertex triangulation of a $(k-1)$-connected $(2k-1)$-pseudomanifold with isolated singularities is tight if and only if*

$$n \cdot \binom{n-k-2}{k} = (-1)^k \binom{2k}{k} \chi(M).$$

7.12 Theorem: *Let P be a d-polytope with N vertices such that each vertex figure of P is simplicial, and let $M \subset P \subset E^d$ be a tight subcomplex which contains each of the N vertices and which is a $(k-1)$-connected $(2k-1)$-pseudomanifold with isolated singularities . Then the following holds*

$$N \cdot \binom{d-k-1}{k} \leq (-1)^k \binom{2k}{k} \chi(M).$$

Moreover, for $d \geq 2k+1$ equality holds if and only if P is simple (i.e., each vertex figure is a simplex, Def. 1.1).

7.13 Corollary: *Under the same assumptions as in 7.12 the following inequality holds*

$$(d+1) \binom{d-k-1}{k} \leq (-1)^k \binom{2k}{k} \chi(M)$$

with equality for $d \geq 2k+1$ if and only if P is a simplex.

Compare 7.9 giving the same type of conclusion but under different assumptions about P.

Conjecture D: *7.9 holds for any tight polyhedral embedding of a $(k-1)$-connected $(2k-1)$-pseudomanifold with isolated singularities into E^d.*

Conjecture E: *7.12 holds for any combinatorial $(2k-1)$-pseudomanifold with isolated singularities.*

These conjectures are analogues of Conjecture A and Conjecture B, respectively, compare Chapter 4.

7.14 Proposition: *Conjecture E is true for a fixed k if Conjecture B is true for $k - 1$. In particular, Conjecture E is true for $k = 2$ and $k = 3$.*

PROOF OF 7.9: This follows the pattern of the proof of 4.5. Again the case $d = 2k$ is trivial, so we assume $d \geq 2k + 1$. We start with the equation from 7.8

$$h_{k+1}(M) - h_{k-1}(M) = (-1)^k \left(\binom{2k-1}{k-1} - \binom{2k-1}{k-2} \right) \chi(M)$$

$$= (-1)^k \cdot \frac{2}{k-1} \binom{2k-1}{k-2} \chi(M)$$

and then put in successively the inqualities from 4.4

$$h_{j+1}(P) - h_j(P) \geq 0.$$

The assumptions on tightness and $(k - 1)$-connectivity imply that M contains $Sk_k(P)$. Therefore the first parts of the f-vectors $(f_{-1}, f_0, f_1, \ldots, f_k)$ of M and P coincide as in the proof of 4.5. Let $c^i_{j,d}$ denote the corresponding coefficients of

$$h_{k+1}(M) - h_{k-1}(M) \;=\; \sum_{i=-1}^{k} (-1)^{k-i} \left(\binom{2k-i}{k-i} - \binom{2k-i}{k-i-1} \right) f_i$$

$$\cdots$$

$$\geq \sum_{i=-1}^{j} (-1)^{j-i} c^i_{j,d} f_i$$

$$\cdots$$

$$\geq c^0_{0,d} f_0 - c^{-1}_{0,d}$$

$$\geq c^{-1}_{-1,d}.$$

We have the following recursion formula

$$c^i_{k,d} = \binom{2k-i}{k} - \binom{2k-i}{k+1} = \frac{i+1}{k+1} \binom{2k-i}{k}$$

$$c^i_{j-1,d} = c^j_{j,d} \binom{d-i}{j-i} - c^i_{j,d} \quad \text{for } j > i.$$

The proof will be completed by the following lemma which in particular says

$$c_{0,d}^0 = \frac{d-1}{k}\binom{d-k-2}{k-1} > 0$$

and

$$c_{-1,d}^{-1} = \frac{d+1}{k+1}\binom{d-k-1}{k}$$

which implies

$$(d+1)\binom{d-k-1}{k} \le (-1)^k \cdot 2\frac{k+1}{k-1}\binom{2k-1}{k-2}\chi(M) = (-1)^k\binom{2k}{k}\chi(M).$$

if $d \ge 2k+1$ then in the case of equality we obtain $f_0 = d+1$. By 7.7 M is a tight triangulation. It also follows that the link of each vertex is a tight triangulation of a $(k-2)$-connected $(2k-2)$-manifold with $d \ge 2k+1$ vertices which is certainly not a sphere by 4.7. Hence each vertex of M is a proper singularity in this case.

LEMMA:

$$c_{i,d}^i = \binom{d-k-i-1}{k-i} + \binom{d-k-i-2}{k-i-1}$$

$$= \begin{cases} 1 & \text{for } i = k \text{ and any } d \\ 2 & \text{for } -1 \le i \le k-1 \text{ and } d = 2k+1 \\ \dfrac{d-2i-1}{d-2k-1}\binom{d-k-i-2}{k-i} & \text{otherwise.} \end{cases}$$

PROOF: The case $c_{k,d}^k = 1$ is clear. In the case $d = 2k+1$ we use induction backwards. First of all we have

$$c_{k-1,2k+1}^{k-1} = d - k + 1 - \left(\binom{k+1}{1} - \binom{k+1}{0}\right) = 2.$$

Now we assume $c_{j,2k+1}^j = 2$ for $j \ge i+1$ and apply the recursion formula as in 4.5:

$$c_{i,2k+1}^i = \sum_{j=i+1}^{k} (-1)^{j-i-1} c_{j,2k+1}^j \binom{d-i}{j-i} + (-1)^{k-i} c_{k,2k+1}^i$$

$$= 2 \sum_{j=i+1}^{k-1} (-1)^{j-i-1} \binom{2k+1-i}{j-i} + (-1)^{k-i-1} \binom{2k+1-i}{k-i} +$$

$$+ (-1)^{k-i} \left(\binom{2k-i}{k} - \binom{2k-i}{k+1} \right)$$

$$= 2.$$

Secondly, we have Pascal's rule

(PR) $c_{i,d}^i + c_{i+1,d+1}^{i+1} = c_{i,d+1}^i$

which holds by exactly the same calculations as in 4.5 because the recursion formula for the $c_{i,d}^i$ remains unchanged. The proof is completed by the observation that the numbers on the right hand side of the lemma obey the same rule (PR). The numbers $c_{i,d}^i$ happen to be the entries of the following modification of Pascal's triangle:

$$i = k \qquad\qquad d = 2k+1$$

$$\searrow \qquad \swarrow$$

					1			
				2		1		
			2		3		1	
		2		5		4		1
	2		7		9		5	1
2		9		16		14	6	1

PROOF OF 7.10: This is essentially the same as for 4.6: We start with the equation

$$h_{k+1}(M) - h_{k-1}(M) = (-1)^k \frac{2}{k-1} \binom{2k-1}{k-2} \chi(M).$$

Then we put in $f_0 = n$, $f_i = \binom{n}{i+1}$ for $i = 1, \ldots, k-1$ and the inequality $f_k \leq \binom{n}{k+1}$. The details are omitted here. In the case of equality $f_k = \binom{n}{k+1}$ the triangulation is $(k+1)$-neighborly and therefore $(k-1)$-tight and $(k-1)$-connected by 3.10. 7.7 implies the tightness.

PROOF OF 7.12: By assumption M is $(k-1)$-connected and tight. 3.5 implies that M is k-Hamiltonian in P. It follows that the link of each vertex v_i is $(k-1)$-Hamiltonian in the vertex figure P_i of P at v_i which by assumption is simplicial. Note that M covers each vertex of P_i by the 0-tightness. Since this link is a $(2k-2)$-manifold it is a tight subcomplex of P_i by 4.1. Therefore we can apply 4.5 to the link L_i of each vertex v_i regarded as a subcomplex of the $(d-1)$-polytope P_i:

$$\binom{d-k-1}{k} \leq (-1)^{k-1}\binom{2k-1}{k}\left(\chi(L_i)-2\right).$$

Summation over $i = 1, \ldots, N$ leads to

$$N\binom{d-k-1}{k} \leq (-1)^k\binom{2k}{k} \cdot \frac{1}{2}\sum_{i=1}^{N}\left(2 - \chi(L_i)\right),$$

and the assertion follows from Lemma 7.15. In the case of equality each P_i is a simplex by 4.5, hence P is simple.

7.15 Lemma: *For any $(2k-1)$-pseudomanifold M with isolated singularities the following equation holds:*

$$2\chi(M) = \sum_i \left(2 - \chi(L_i)\right)$$

where L_i denotes the link of the ith vertex, and where the sum ranges over all vertices.

PROOF: Let $\overline{M}$ denote the $(2k-1)$-manifold with boundary which we get by cutting out the singularities of M. Furthermore let $2\overline{M}$ be the usual doubling of $\overline{M}$ defined as $2\overline{M} := \overline{M} \cup_{\partial\overline{M}} \overline{M}$. By construction $\partial\overline{M} \cong \bigcup_i L_i$, and the assertion follows from the additivity formula

$$\chi(M) + \chi(\partial\overline{M}) = \chi(\overline{M}) + \sum_i 1$$

and from

$$0 = \chi(2\overline{M}) = 2\chi(\overline{M}) - \chi(\partial\overline{M}).$$

PROOF OF 7.14: Let M be a combinatorial $(2k-1)$-pseudomanifold with n vertices. Each of the vertex links $L_1, \ldots, L_n$ has $n_i \leq n - 1$ vertices. By assumption Conjecture B holds for each L_i, i.e.

$$\binom{n_i - k - 1}{k} \geq (-1)^{k-1} \binom{2k-1}{k} (\chi(L_i) - 2)$$

with equality if and only if L_i is k-neighborly. The assertion follows by summation over all i and by Lemma 7.15.

REMARK: For the validity of Conjecture E for a particular pseudomanifold M it is, of course, sufficient that Conjecture B holds for each vertex link of M. If M is a $(2k-1)$-manifold then either one is trivially satisfied.

7.16 Examples:

1. *There are tight triangulations of certain simply connected 3-pseudomanifolds with $8, 17, 29$ vertices. These satisfy equality in $7.9, 7.10$ and 7.11.*

2. *There are infinitely many 3-dimensional examples satisfying equality in 7.12, and there are also higher dimensional examples.*

The construction in case 2 is just $2^K \subset C^d$ where K is a tight triangulation of a $(k-2)$-connected $(2k-2)$-manifold with d vertices. There are infinitely many examples for $k = 2$, see 2.16. Higher dimensional examples for K are given in 4.13 and 4.17. In each case the polytope is the cube C^d with $N = 2^d$ vertices. The Euler characteristic in this case satisfies (by 4.7 and 7.12)

$$\chi(2^K) = (-1)^k \cdot 2^d \binom{d-k-1}{k} \binom{2k}{k}^{-1} = 2^{d-1}(2 - \chi(K)).$$

Equivalently, the Betti numbers satisfy

$$b_k(2^K) = 2^{d-1} b_{k-1}(K).$$

Examples for case 1 can be found in the literature. As a twofold quadruple system, a special type of a block design, A. Emch [Em;p.39] described a scheme of 28 quadruples on 8 elements which form a 3-neighborly combinatorial 3-pseudomanifold M_8 with $n = 8$ vertices. This structure is called $\mathcal{T}$ in [Al4]), its dual has been found independently by B. Grünbaum as a so-called polystroma. The link of each vertex is the 7-vertex triangulation of the torus (Figure 2). Regarding the 7 vertices as elements of $\mathbf{Z}_7$ the full affine group in dimension one over $\mathbf{Z}_7$ acts on the triangulation (compare [Kü-L2]). It is just the orbit of the triangle $\langle 013 \rangle$ under the affine group over $\mathbf{Z}_7$. To construct M_8 we take the orbit of the tetrahedron $\langle 013x \rangle$ under the action of the full projective group PSL(2,7) of order 168 where x denotes the point at infinity of the projective line PG(1,7). Under the group of translations in $\mathbf{Z}_7$

there are four orbits generated by $\langle 013x \rangle$, $\langle 023x \rangle$, $\langle 2456 \rangle$, $\langle 1456 \rangle$. This generates 28 tetrahedra satisfying in addition a complementarity condition: for each tetrahedron $\langle v_1 v_2 v_3 v_4 \rangle$ the complementary tetrahedron $\langle v_5 v_6 v_7 v_8 \rangle$ is also in the complex where $\{v_1, \ldots, v_8\} = \{1, \ldots, 7, x\}$. Its Euler characteristic is $\chi = 8 - \binom{8}{2} + \binom{8}{3} - 28 = 8$ in agreement with 7.11: $8 \cdot \binom{4}{2} = \binom{4}{2} \cdot \chi$.

The two examples with 17 and 29 vertices are due to E. Köhler [Kö]. For $n = 17$ we regard the vertices as elements of $\mathbf{Z}_{17}$. The pseudomanifold M_{17} is the union of the orbits of

$$\langle 0145 \rangle \quad \text{and} \quad \langle 0148 \rangle$$

under the group of affine transformations $x \mapsto ax + b$ of $\mathbf{Z}_{17}$. This yields $20 \cdot 17 = 340$ tetrahedra, and the automorphism group acts transitively on the set of vertices and edges. The link of the edge $\langle 01 \rangle$ turns out to be the closed circuit

$$(3\ 6\ 2\ 13\ 14\ 10\ 7\ 11\ 8\ 4\ 5\ 16\ 12\ 15\ 9)$$

of length 15.

For $n = 29$ we start with $\mathbf{Z}_{29}$ and define M_{29} as the union of the orbits of

$$\langle 0\ 1\ 12\ 13 \rangle, \langle 0\ 1\ 12\ 24 \rangle, \langle 0\ 1\ 11\ 23 \rangle$$

under the group of affine transformations of $\mathbf{Z}_{29}$. This yields $63 \cdot 29 = 1827$ tetrahedra. The link of the edge $\langle 01 \rangle$ is the closed circuit

$$(7\ 19\ 5\ 3\ 24\ 12\ 13\ 28\ 16\ 4\ 22\ 20\ 9\ 21\ 10\ 8\ 26\ 14\ 2\ 17\ 18\ 6\ 27\ 25\ 11\ 23\ 15)$$

of length 27.

Compare the Euler characteristics $\chi(M_{17}) = 221$, $\chi(M_{29}) = 1450$ satisfying the equation in 7.11.

REMARK: In terms of block designs any 3-neighborly 3-pseudomanifolds with n vertices is a design $\bar{S}_2(3, 4; n)$, compare [BJL]. On the other hand a given $S_2(3, 4; n)$ can be interpreted as a 3-pseudomanifold if the link of each vertex is a triangulated surface without singularities. It is quite possible that the other examples given in [Kö] satisfy this condition, in particular those with a prime number of vertices $n \equiv 5 \mod 12$. We have not checked the details.

7.17 Proposition (Tight slicings): *Let M be a subcomplex of a convex d-polytope P such that $P = \mathcal{H}(M)$ and such that M is a k-connected $(2k + 1)$-pseudomanifold with isolated singularities. Then the following conditions are equivalent.*

(i) M is tight in E^d.

(ii) $M \cap E^{d-1}$ is tight in E^{d-1} for every hyperplane $E^{d-1} \subset E^d$ which does not meet any vertex of M.

Note that for such an E^{d-1} the intersection $M \cap E^{d-1}$ is a $(k - 1)$-connected $2k$-manifold (without singularities). We call it a <u>slicing of M</u>.

PROOF: By 7.6 M satisfies (i) if and only if M is $(k + 1)$-Hamiltonian in P. This implies that each slicing $M \cap E^{d-1}$ is k-Hamiltonian in $P \cap E^{d-1}$ and thus tight by 4.1. Conversely, if (ii) is satisfied then $M \cap E^{d-1}$ is k-Hamiltonian in $P \cap E^{d-1}$ for every such E^{d-1}, and this implies that M is $(k + 1)$-Hamiltonian in P because each $(k + 1)$-face is the union of its k-dimensional slicings.

7.17 gives another method for constructing tight polyhedral $(k - 1)$-connected $2k$-manifolds, namely as slicings. The method of truncating and fitting in tight triangulations (see 4.16, 6.11) leads to similar results but the geometry of the examples is different in general. The method in 7.17 does not require any tight triangulation. Note that in general slicings of tight polyhedra are not tight. The slicing of a tight surface in E^d yields closed curves in E^{d-1} which cannot be tight for $d \geq 4$.

In order to illustrate 7.17 we consider the smallest example in 7.16 which is a tight 8-vertex pseudomanifold $M_8 \subset \triangle^7$. Regarding $\triangle^7$ as the join $\triangle^7 = \triangle_1^3 * \triangle_2^3$ where $\triangle_1^3 = \langle 0124 \rangle$ and $\triangle_2^3 = \langle 365x \rangle$, we take a hyperplane E^6 half way between $\triangle_1^3$ and $\triangle_2^3$ (which happen to be tetrahedra of M_8). The intersection $M_8 \cap E^6$ is a surface of genus 4 which is invariant under the automorphism $S = (124)(365)$ and also under $R = (0x)(16)(23)(45)$, and it is 1-Hamiltonian and hence tight in the 6-polytope $\triangle^7 \cap E^6$. Compare the hexagonal slice of an ordinary 3-cube half way between four vertices on either side. Figure 11 shows the combinatorial structure of $M_8 \cap E^6$ where S corresponds to rotation by $2\pi/3$ and where 25 denotes the centre of the edge $\langle 25 \rangle$ etc., e.g., $0x$ is the only vertex fixed under S. The fundamental domain of the surface is indicated by broken lines in Figure 11. The shaded triangles are not contained but rather induce handles attached to a torus of hexagonal shape.

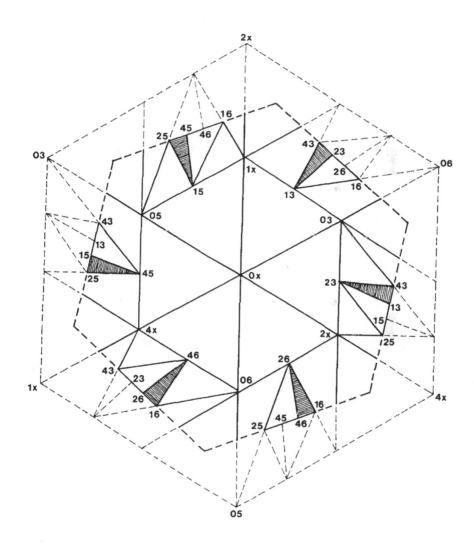

Figure 11

References

[A] A. D. ALEXANDROV, On a class of closed surfaces, *Recueil Math.(Moscow)* **4**, 69 - 72 (1938)

[Al1] A. ALTSHULER, Polyhedral realization in $\mathbf{R}^3$ of triangulations of the torus and 2-manifolds in cyclic 4-polytopes, *Discr. Math.* **1**, 211 - 238 (1971)

[Al2] —, Combinatorial 3-manifolds with few vertices, *J. Combin. Th. (A)* **16**, 165 - 173 (1974)

[Al3] —, Neighborly 4-polytopes and neighborly combinatorial 3-manifolds with ten vertices, *Can. J. Math.* **29**, 400 - 420 (1977)

[Al4] —, 3-pseudomanifolds with preassigned links, *Trans. Amer. Math. Soc.* **241**, 213 - 237 (1978)

[ABS1] A. ALTSHULER, J. BOKOWSKI and J. SCHUCHERT, Spatial polyhedra without diagonals, *Israel J. Math.* **86**, 373 - 396 (1994)

[ABS2] —, Neighborly 2-manifolds with 12 vertices, preprint 1995

[AB] A. ALTSHULER and U. BREHM, Neighborly maps with few vertices, *Discr. Comp. Geom.* **8**, 93 - 104 (1992)

[AS1] A. ALTSHULER and L. STEINBERG, Neighborly combinatorial 3-manifolds with 9 vertices, *Discr. Math.* **8**, 113 - 137 (1974)

[AS2] —, An enumeration of combinatorial 3-manifolds with nine vertices, *Discr. Math.* **16**, 91 - 108 (1976)

[AM] P. ARNOUX and A. MARIN, The Kühnel triangulation of the complex projective plane from the view-point of complex crystallography, Part II, *Memoirs Fac. Sci. Kyushu Univ., Ser. A* **45**, 167 - 244 (1991)

[BD] B. BAGCHI and B. DATTA, On Kühnel's 9-vertex complex projective plane, *Geom. Dedicata* **50**, 1 - 13 (1994)

[Ba1] T. F. BANCHOFF, Tightly embedded 2-dimensional polyhedral manifolds, *Amer. J. Math.* **87**, 462 - 472 (1965)

[Ba2] —, Critical points and curvature for embedded polyhedra, *J. Diff. Geom.* **1**, 245 - 256 (1967)

[Ba3] —, Non-rigidity theorems for tight polyhedra, *Arch. Math.* **21**, 416 - 423 (1970)

[Ba4] —, High codimensional 0-tight maps on spheres, *Proc. Amer. Math. Soc.* **29**, 133 - 137 (1971)

[Ba5] —, Tight polyhedral Klein bottles, projective planes, and Möbius bands, *Math. Ann.* **207**, 233 - 243 (1974)

[Ba6] —, Critical points and curvature for embedded polyhedra II, *Progress in Math.* **32**, 32 - 55 (1983)

[Ba7] —, Normal curvatures and Euler class for polyhedral surfaces in 4-space, *Proc. Amer. Math. Soc.* **92**, 593 - 596 (1984)

[Ba-Kü] T. F. BANCHOFF and W. KÜHNEL, Equilibrium triangulations of the complex projective plane, *Geom. Dedicata* **44**, 313 - 333 (1992)

[Ba-Kui] T. F. BANCHOFF and N. H. KUIPER, Geometrical class and degree for surfaces in three-space, *J. Diff. Geom.* **16**, 559 - 576 (1981)

[Ba-T] T. F. BANCHOFF and F. TAKENS, Height functions on surfaces with three critical points, *Illinois J. Math.* **19**, 325 - 335 (1975)

[Bar1] D. BARNETTE, A proof of the lower bound conjecture for convex polytopes, *Pac. J. Math.* **46**, 349 - 354 (1973)

[Bar2] —, The triangulations of the 3-sphere with up to 8 vertices, *J. Combin. Th. (A)*, **14**, 37 - 52 (1973)

[Bar-G] D. BARNETTE and D. GANNON, Manifolds with few vertices, *Discr. Math.* **16**, 291 - 298 (1976)

[BH] L. W. BEINEKE and F. HARARY, The genus of the n-cube, *Canad. J. Math.* **17**, 494 - 496 (1965)

[BJL] T. BETH, D. JUNGNICKEL and H. LENZ, *Design Theory*, B.I. Mannheim-Wien-Zürich 1985

[Bj-K] A. BJÖRNER and G. KALAI, On f-vectors and homology, *Combinatorial Mathematics* (G.J. Bloom et al., eds.) *Proc. New York Acad. Sci.* **555**, 63 - 80 (1989)

[Bo-Eg] J. BOKOWSKI and A. EGGERT, All realizations of Möbius' torus with 7 vertices, *Structural Topology* **17**, 59 - 76 (1991)

[Bo-Br] J. BOKOWSKI and U. BREHM, A polyhedron of genus 4 with minimal number of vertices and maximal symmetry, *Geom. Dedicata* **29**, 53 - 64 (1989)

112

[Br1] U. BREHM, Polyeder mit zehn Ecken vom Geschlecht drei, *Geom. Dedicata* **11**, 119 - 124 (1981)

[Br2] —, Non-embeddable triangulations, 3. Koll. Diskrete Geometrie, Salzburg 1985

[Br3] —, A minimal polyhedral Dirichlet tesselation of the complex projective plane, in preparation

[Br4] —, The homology of the power-complex of a simplicial complex, in preparation

[Br-Kü1] U. BREHM and W. KÜHNEL, Smooth approximation of polyhedral surfaces regarding curvatures, *Geom. Dedicata* **12**, 435 - 461 (1982)

[Br-Kü2] —, A polyhedral model for Cartan's hypersurface in S^4, *Mathematika* **33**, 55 - 61 (1986)

[Br-Kü3] —, Combinatorial manifolds with few vertices, *Topology* **26**, 465 - 473 (1987)

[Br-Kü4] —, 15-vertex triangulations of an 8-manifold, *Math. Ann.* **294**, 167 - 193 (1992)

[Br-Sw] U. BREHM and J. ŚWIATKOWSKI, Triangulations of lens spaces with few simplices, preprint TU Berlin 1993

[Bu-Kü] P. BREUER and W. KÜHNEL, The tightness of tubes, preprint 1995

[CR] T. E. CECIL and P. J. RYAN, *Tight and Taut Immersions of Manifolds*, Pitman, Boston-London-Melbourne 1985

[Ce1] D. CERVONE, Tight immersions of simplicial surfaces into three-space, *Topology* (to appear)

[Ce2] —, A tight polyhedral immersion of the real projective plane with one handle, in preparation; see the URL http://www.geom.umn.edu/docs/dpvc/RP2.html

[Ch] J. CHEEGER, A vanishing theorem for piecewise constant curvature spaces, *Curvature and Topology of Riemannian Manifolds*, (K. Shiohama et al., eds.), Proc. Katata 1985, 33 - 40, Lecture Notes in Mathematics **1201**, Springer , Berlin-Heidelberg-New York 1986

[CL] S. S. CHERN and R. K. LASHOF, On the total curvature of immersed manifolds, I, II, *Amer. J. Math.* **79**, 306 - 318 (1957), *Mich. Math. J.* **5**, 5 - 12 (1958)

[CCHR] C. J. COLBOURN, M. J. COLBOURN, J. J. HARMS, A. ROSA, A complete census of (10,3,2) block designs and of Mendelsohn triple systems of order ten III. (10,3,2) block designs without repeated blocks, *Congressus Numerantium* **37**, 211 - 234 (1983)

[Con1] R. CONNELLY, A flexible sphere, *The Math. Intelligencer* **1**, 130 - 131 (1978)

[Con2] —, The rigidity of certain cabled frameworks and the second-order rigidity of arbitrarily triangulated convex surfaces, *Adv. Math.* **37**, 272 - 299 (1980)

[Cox1] H. S. M. COXETER, Regular skew polyhedra in three and four dimensions, and their topological analogues, *Proc. London Math. Soc. (2)* **43**, 33 - 62 (1937), reprinted in: *Twelve Geometric Essays*, 76 - 105, Southern Illinois University Press 1968

[Cox2] —, Twisted honeycombs, *Regional Conference Series in Math.*, No. 4, Amer. Math. Soc., Providence 1970

[Cox-M] H. S. M. COXETER and W. O. J. MOSER, *Generators and Relations for Discrete Groups*, 4th ed., Springer, Berlin-Heidelberg-New York 1980 (*Ergebnisse der Mathematik und ihrer Grenzgebiete* 14)

[Cs] A. CSÁSZÁR, A polyhedron without diagonals, *Acta Sci. Math. (Szeged)* **13**, 140 - 142 (1949)

[Dc] J. DANCIS, Triangulated n-manifolds are determined by their $[n/2] + 1$-skeletons, *Topology and its applications* **18**, 17 - 26 (1984)

[Dz] L. DANZER, Regular incidence-complexes and dimensionally unbounded sequences of such, I: *Annals Discrete Math.* **20**, 115 - 127 (1984), II: in preparation

[Do] A. DOLD, *Lectures on Algebraic Topology*, Springer, Berlin-Heidelberg-New York 1972 (*Die Grundlehren der mathematischen Wissenschaften in Einzeldarstellungen* 200)

[Du] R. A. DUKE, Geometric embedding of complexes, *Amer. Math. Monthly* **77**, 597 - 603 (1970)

[Ed] R. D. EDWARDS, The double suspension of a certain homology 3-sphere is S^5, *Notices Amer. Math. Soc.* **22**, A - 334 (1975)

[EK] J. EELLS and N. H. KUIPER, Manifolds which are like projective planes, *Publ. Math. I.H.E.S.* **14**, 181 - 222 (1962)

[Em] A. EMCH, Triple and multiple systems, their geometric configurations and groups, *Trans. Amer. Math. Soc.* **31**, 25 - 42 (1929)

[Fe] D. FERUS, *Totale Absolutkrümmung in Differentialgeometrie und -topologie*, Lecture Notes in Mathematics 66, Springer, Berlin-Heidelberg-New York 1968

[Fr] P. FRANKLIN, A six colour problem, *J. of Mathematics and Physics* **13**, 363 - 369 (1934)

114

[Gri] P. GRITZMANN, Tight polyhedral realisations of closed 2-dimensional manifolds in $\mathbb{R}^3$, *J. Geom.* **17**, 69 - 76 (1981)

[Grü] B. GRÜNBAUM, *Convex Polytopes*, Interscience Publishers, New York 1967

[Grü-S] B. GRÜNBAUM and V. P. SREEDHARAN, An enumeration of simplicial 4-polytopes with 8 vertices, *J. Comb. Th.* **2**, 437 - 465 (1967)

[Haa] F. HAAB, Immersions tendues de surfaces dans E^3, *Comment. Math. Helv.* **67**, 182 - 202 (1992)

[Har] J. B. HARTLE, Simplicial minisuperspace I. General discussion, II. Some classical solutions, *J. Math. Phys.* **26**, 804 - 814 (1985) and **27**, 287 - 295 (1986)

[Hea] P. J. HEAWOOD, Map colour theorem, *Quart. J. Math.* **24**, 332 - 338 (1890)

[Hud] J. F. P. HUDSON, *Piecewise Linear Topology*, University of Chicago Lecture Notes, W.A. Benjamin, New York 1969

[Hun] J. P. HUNEKE, A minimum-vertex triangulation, *J. Combin. Th. (B)* **24**, 258 - 266 (1978)

[J] M. JUNGERMAN, The non-orientable genus of the n-cube, *Pacific J. Math.* **76**, 443 - 451 (1978)

[J-Ri] M. JUNGERMAN and G. RINGEL, Minimal triangulations on orientable surfaces, *Acta Math.* **145**, 121 - 154 (1980)

[J-Ri1] —, The genus of the n-octahedron: Regular cases, *J. Graph Th.* **2**, 69 - 75 (1978)

[Kal1] G. KALAI, Rigidity and the Lower Bound Theorem I, *Invent. Math.* **88**, 125 - 151 (1987)

[Kal2] —, Many triangulated spheres, *Discr. Comp. Geom.* **3**, 1 - 14 (1988)

[Kal3] —, The combinatorial theory of convex polytopes, *Polytopes: Abstract, Convex and Computational* (T. Bisztriczky et al., eds.), NATO Adv. Study Inst. Ser. C, Math. Phys. Sci. **440**, 205 - 229, Kluwer, Dordrecht 1994

[Ki] R. KIRBY, *The Topology of 4-Manifolds*, Lecture Notes in Mathematics **1374**, Springer, Berlin-Heidelberg-New York etc. 1989

[Kl1] V. KLEE, A combinatorial analogue of Poincaré's duality theorem, *Can. J. Math.* **16**, 517 - 531 (1964)

[Kl2] —, The number of vertices of a convex polytope, *Can. J. Math.* **16**, 701 - 720 (1964)

[Kö] E. KÖHLER, Quadruple systems over Z_p admitting the affine group, *Combinatorial Theory* (D. Jungnickel and K. Vedder, eds.), 212 - 228, Lecture Notes in Mathematics **969**, Springer, Berlin-Heidelberg-New York 1982

[Kü1] W. KÜHNEL, Total absolute curvature of polyhedral manifolds with boundary in E^n, *Geom. Dedicata* **8**, 1 - 12 (1979)

[Kü2] —, Tight and 0-tight polyhedral embeddings of surfaces, *Invent. Math.* **58**, 161 - 177 (1980)

[Kü3] —, Higherdimensional analogues of Császár's torus, *Res. Math.* **9**, 95 - 106 (1986)

[Kü4] —, Minimal triangulations of Kummer varieties, *Abh. Math. Sem. Univ. Hamburg* **57**, 7 - 20 (1987)

[Kü5] —, Triangulations of manifolds with few vertices, *Advances in Differential Geometry and Topology* (F. Tricerri, ed.), 59 - 114, World Scientific, Singapore 1990

[Kü6] —, Tightness, torsion, and tubes, *Ann. Glob. Analysis Geom.* **10**, 227 - 236 (1992)

[Kü7] —, Hamiltonian surfaces in polytopes, *Intuitive Geometry*, Proc. Conf. Szeged 1991 (K. Böröczky et al., eds.), 197 - 203, Coll. Math. Soc. J. Bolyai **63**, North-Holland, Amsterdam 1994

[Kü8] —, Manifolds in the skeletons of convex polytopes, tightness, and generalized Heawood inequalities, *Polytopes: Abstract, Convex and Computational* (T. Bisztriczky et al., eds.), NATO Adv. Study Inst. Ser. C, Math. Phys. Sci. **440**, 241 - 247, Kluwer, Dordrecht 1994

[Kü9] —, Tensor products of spheres, *Geometry and Topology of Submanifolds, VI* (F. Dillen et al., eds.), 106 - 109, World Scientific, Singapore 1994

[Kü-Ba] W. KÜHNEL and T. F. BANCHOFF, The 9-vertex complex projective plane, *The Math. Intelligencer* Vol. 5, issue 3, 11 - 22 (1983)

[Kü-L1] W. KÜHNEL and G. LASSMANN, The unique 3-neighborly 4-manifold with few vertices, *J. Combin. Th. (A)* **35**, 173 - 184 (1983)

[Kü-L2] —, The rhombidodecahedral tessellation of 3-space and a particular 15-vertex triangulation of the 3-dimensional torus, *manuscripta math.* **49**, 61 - 77 (1984)

[Kü-L3] —, Neighborly combinatorial 3-manifolds with dihedral automorphism group, *Israel J. Math.* **52**, 147 - 166 (1985)

[Kü-Pi] W. KÜHNEL and U. PINKALL, Tight smoothing of some polyhedral surfaces, *Global Differential Geometry and Global Analysis* (D. Ferus et al., eds.), Proceedings Berlin 1984, 227 - 239, Lecture Notes in Mathematics **1156**, Springer, Berlin-Heidelberg-New York 1985

[Kü-Sch] W. Kühnel and Ch. Schulz, Submanifolds of the cube, *Applied Geometry and Discrete Mathematics, The Victor Klee Festschrift* (P. Gritzmann and B. Sturmfels, eds.), 423 - 432, DIMACS Series in Discrete Math. and Theor. Comp. Sci. **4**, Amer. Math. Soc. 1991

[Kui1] N. H. Kuiper, Immersions with minimal total absolute curvature, Colloque de Géometrie différentielle globale du Centre Belge de Recherches Math., Bruxelles 1958, 75 - 88 (1958)

[Kui2] —, Convex immersions of closed surfaces in E^3, *Comment. Math. Helv.* **35**, 85 - 92 (1961)

[Kui3] —, On convex maps, *Nieuw Archief voor Wiskunde (3)* **10**, 147 - 164 (1962)

[Kui4] —, Minimal total absolute curvature for immersions, *Invent. Math.* **10**, 209 - 238 (1970)

[Kui5] —, Morse relations for curvature and tightness, *Proc. Liverpool Singularities Symp. II* (C.T.C. Wall, ed.), 77 - 89, Lecture Notes in Mathematics **209**, Springer, Berlin-Heidelberg-New York 1971

[Kui6] —, Tight topological embeddings of the Möbius band, *J. Diff. Geom.* **6**, 271 - 283 (1972)

[Kui7] —, A short history of triangulation and related matters, Proc. bicenten. Congr. Wiskd. Genoot., Part I, Amsterdam 1978, Math. Cent. Tracts **100**, 61 - 79 (1979)

[Kui8] —, Tight embeddings and maps. Submanifolds of geometrical class three in E^N, *The Chern Symposium* (W.-Y. Hsiang et al., eds.), Berkeley 1979, 97 - 145, Springer, New York-Heidelberg-Berlin 1980

[Kui9] —, Geometry in total absolute curvature theory, *Perspectives in Mathematics*, Anniversary of Oberwolfach 1984 (W. Jäger et al., eds.), 377 - 392, Birkhäuser, Basel-Boston-Stuttgart 1984

[Kui10] —, There is no tight continuous immersion of the Klein bottle into $\mathbf{R}^3$, preprint I.H.E.S. 1983

[Kui-P] N. H. Kuiper and W. F. Pohl, Tight topological embeddings of the real projective plane in E^5, *Invent. Math.* **42**, 177 - 199 (1977)

[LR] N. Levitt and C. Rourke, The existence of combinatorial formulae for characteristic classes, *Trans. Amer. Math. Soc.* **239**, 391 - 397 (1978)

[M] P. Mani, Spheres with few vertices, *J. Comb. Th. (A)* **13**, 346 - 352 (1972)

[Ma] G. Mannoury, Surfaces-images, *Nieuw Archief voor Wiskunde* **4**, 112 - 129 (1900)

[MR] R. MATHON and A. ROSA, A census of Mendelsohn triple systems of order nine, *Ars Comb.* **4**, 309 - 315 (1977)

[MM] P. MCMULLEN, On simple polytopes, *Invent. Math.* **113**, 419 - 444 (1993)

[MM-S] P. MCMULLEN and E. SCHULTE, Regular polytopes from twisted Coxeter groups, *Math. Z.* **201**, 209 - 226 (1989)

[MM-Sh] P. MCMULLEN and G. C. SHEPHARD, *Convex Polytopes and the Upper Bound Conjecture*, London Mathematical Society Lecture Note Series **3**, Cambridge University Press 1971

[MM-Wa] P. MCMULLEN and D. WALKUP, A generalised lower bound conjecture for simplicial polytopes, *Mathematika* **18**, 264 - 273 (1971)

[MP] R. MACPHERSON, The combinatorial formula of Gabrielov, Gelfand and Losik for the first Pontrjagin class, *Séminaire Bourbaki, 29e année*, no. 497, 1977, Lecture Notes in Mathematics **677**, Springer, Berlin-Heidelberg-New York 1978

[Mi] L. MILIN, A combinatorial computation of the first Pontryagin class of the complex projective plane, *Geom. Dedicata* **49**, 253 - 291 (1994)

[Mir] J. MILNOR, On the relationship between the Betti numbers of a hypersurface and an integral of its Gaussian curvature (1950), in: Collected papers, Vol. I, 15 - 26, Publish or Perish 1994

[Mö] A. MÖBIUS, Zur Theorie der Polyëder und der Elementarverwandtschaft, *Gesammelte Werke, Vol.* 2 (F. Klein, ed.), 519 - 559, Hirzel, Leipzig 1886

[MY] B. MORIN and M. YOSHIDA, The Kühnel triangulation of the complex projective plane from the view point of complex crystallography, Part I, *Memoirs Fac. Sci. Kyushu Univ., Ser. A*, **45**, 55 - 142 (1991)

[Oz] T. OZAWA, Products of tight continuous functions, *Geom. Dedicata* **14**, 209 - 213 (1983)

[Pi1] U. PINKALL, Tight surfaces and regular homotopy, *Topology* **25**, 475 - 481 (1986)

[Pi2] —, Curvature properties of taut submanifolds, *Geom. Dedicata* **20**, 79 - 83 (1986)

[Po] W. F. POHL, Tight topological immersions of surfaces in higher dimensions, manuscript 1981

[Ri1] G. RINGEL, Wie man die geschlossenen nichtorientierbaren Flächen in möglichst wenig Dreiecke zerlegen kann, *Math. Ann.* **130**, 317 - 326 (1955)

[Ri2] —, Über drei Probleme am n-dimensionalen Würfel und Würfelgitter, *Abh. Math. Sem. Univ. Hamburg* **20**, 10 - 19 (1955)

[Ri3]   —, *Map Color Theorem*, Springer, Berlin-Heidelberg-New York 1974 (*Die Grundlehren der mathematischen Wissenschaften in Einzeldarstellungen* 209)

[Rn1]   R. RIORDAN, *Combinatorial Identities*, Wiley and Sons, New York 1968

[Rn2]   —, The number of faces of simplicial polytopes, *J. Comb. Th.* **1**, 82 - 95 (1966)

[Ro]   L. L. RODRIGUEZ, *The two-piece-property and relative tightness for surfaces with boundary*, Thesis Brown Univ. 1973

[RS]   C. P. ROURKE and B. J. SANDERSON, *Introduction to Piecewise-Linear Topology*, Springer, Berlin-Heidelberg-New York 1972 (*Ergebnisse der Mathematik und ihrer Grenzgebiete* 69)

[Sar1] K. S. SARKARIA, On neighborly triangulations, *Trans. Amer. Math. Soc.* **227**, 213 - 239 (1983)

[Sar2] —, Heawood inequalities, *J. Comb. Th. (A)* **46**, 50 - 78 (1987)

[Scht] R. SCHULTZ, Some recent results on topological manifolds, *Amer. Math. Monthly* **78**, 941 - 952 (1971); reprinted with additional references in: *Selected papers on Geometry* (A.K. Stehney et al., eds.), The Math. Ass. of America 1979

[Sch1] CH. SCHULZ, Hamilton-Flächen auf Prismen, *Geom. Dedicata* **6**, 267 - 274 (1977)

[Sch2] —, Geschlossene Flächen im Rand des Würfels, *Abh. Math. Sem. Univ. Hamburg* **50**, 89 - 94 (1980)

[Sta1] R. STANLEY, The upper bound conjecture and Cohen-Macaulay-rings, *Studies in Appl. Math.* **54**, 135 - 142 (1975)

[Sta2] —, The number of faces of a simplicial convex polytope, *Adv. Math.* **35**, 236 - 238 (1980)

[Sta3] —, The number of faces of simplicial polytopes and spheres, *Discrete Geometry and Convexity* (J.E. Goodman et al., eds.), *Ann. New York Acad. Sci.* **440**, 212 - 223 (1985)

[Sta4] —, On the number of faces of centrally-symmetric simplicial polytopes, *Graphs and Combinatorics* **4**, 55 - 66 (1987)

[Ste]   N. E. STEENROD, The classification of sphere bundles, *Ann. Math.* **45**, 294 - 311 (1944)

[Th1]   G. THORBERGSSON, Tight immersions of highly connected manifolds, *Comment. Math. Helv.* **61**, 102 - 121 (1986)

[Th2] —, Homogeneous spaces without taut embeddings, *Duke Math. J.* **57**, 347 - 355 (1988)

[Tr] M. D. TRETKOFF, The Fermat surface and its periods, *Recent Developments in Several Complex Variables*, Proc. Princeton 1979, 413 - 428, Princeton Univ. Press 1981

[Tu] W. T. TUTTE, Convex representations of graphs, *Proc. London Math. Soc.* **10**, 304 - 320 (1960)

[Wa] D. WALKUP, The lower bound conjecture for 3- and 4-manifolds, *Acta Math.* **125**, 75 - 107 (1970)

[W] C. T. C. WALL, Classification of $(n - 1)$-connected $2n$-manifolds, *Ann. Math.* **75**, 163 - 189 (1962)

[Wh1] A. T. WHITE, *Graphs, Groups and Surfaces*, North-Holland, Amsterdam 1973

[Wh2] —, Block designs and graph embeddings, *J. Comb. Th. (B)* **25**, 166 - 183 (1978)

[Wht] J. H. C. WHITEHEAD, On simply connected 4-dimensional polyhedra, *Comment. Math. Helv.* **22**, 48 - 92 (1949)

Index

Vol. 1566: B. Edixhoven, J.-H. Evertse (Eds.), Diophantine Approximation and Abelian Varieties. XIII, 127 pages. 1993.

Vol. 1567: R. L. Dobrushin, S. Kusuoka, Statistical Mechanics and Fractals. VII, 98 pages. 1993.

Vol. 1568: F. Weisz, Martingale Hardy Spaces and their Application in Fourier Analysis. VIII, 217 pages. 1994.

Vol. 1569: V. Totik, Weighted Approximation with Varying Weight. VI, 117 pages. 1994.

Vol. 1570: R. deLaubenfels, Existence Families, Functional Calculi and Evolution Equations. XV, 234 pages. 1994.

Vol. 1571: S. Yu. Pilyugin, The Space of Dynamical Systems with the C^0-Topology. X, 188 pages. 1994.

Vol. 1572: L. Göttsche, Hilbert Schemes of Zero-Dimensional Subschemes of Smooth Varieties. IX, 196 pages. 1994.

Vol. 1573: V. P. Havin, N. K. Nikolski (Eds.), Linear and Complex Analysis – Problem Book 3 – Part I. XXII, 489 pages. 1994.

Vol. 1574: V. P. Havin, N. K. Nikolski (Eds.), Linear and Complex Analysis – Problem Book 3 – Part II. XXII, 507 pages. 1994.

Vol. 1575: M. Mitrea, Clifford Wavelets, Singular Integrals, and Hardy Spaces. XI, 116 pages. 1994.

Vol. 1576: K. Kitahara, Spaces of Approximating Functions with Haar-Like Conditions. X, 110 pages. 1994.

Vol. 1577: N. Obata, White Noise Calculus and Fock Space. X, 183 pages. 1994.

Vol. 1578: J. Bernstein, V. Lunts, Equivariant Sheaves and Functors. V, 139 pages. 1994.

Vol. 1579: N. Kazamaki, Continuous Exponential Martingales and *BMO*. VII, 91 pages. 1994.

Vol. 1580: M. Milman, Extrapolation and Optimal Decompositions with Applications to Analysis. XI, 161 pages. 1994.

Vol. 1581: D. Bakry, R. D. Gill, S. A. Molchanov, Lectures on Probability Theory. Editor: P. Bernard. VIII, 420 pages. 1994.

Vol. 1582: W. Balser, From Divergent Power Series to Analytic Functions. X, 108 pages. 1994.

Vol. 1583: J. Azéma, P. A. Meyer, M. Yor (Eds.), Séminaire de Probabilités XXVIII. VI, 334 pages. 1994.

Vol. 1584: M. Brokate, N. Kenmochi, I. Müller, J. F. Rodriguez, C. Verdi, Phase Transitions and Hysteresis. Montecatini Terme, 1993. Editor: A. Visintin. VII. 291 pages. 1994.

Vol. 1585: G. Frey (Ed.), On Artin's Conjecture for Odd 2-dimensional Representations. VIII, 148 pages. 1994.

Vol. 1586: R. Nillsen, Difference Spaces and Invariant Linear Forms. XII, 186 pages. 1994.

Vol. 1587: N. Xi, Representations of Affine Hecke Algebras. VIII, 137 pages. 1994.

Vol. 1588: C. Scheiderer, Real and Étale Cohomology. XXIV, 273 pages. 1994.

Vol. 1589: J. Bellissard, M. Degli Esposti, G. Forni, S. Graffi, S. Isola, J. N. Mather, Transition to Chaos in Classical and Quantum Mechanics. Montecatini Terme, 1991. Editor: S. Graffi. VII, 192 pages. 1994.

Vol. 1590: P. M. Soardi, Potential Theory on Infinite Networks. VIII, 187 pages. 1994.

Vol. 1591: M. Abate, G. Patrizio, Finsler Metrics – A Global Approach. IX, 180 pages. 1994.

Vol. 1592: K. W. Breitung, Asymptotic Approximations for Probability Integrals. IX, 146 pages. 1994.

Vol. 1593: J. Jorgenson & S. Lang, D. Goldfeld, Explicit Formulas for Regularized Products and Series. VIII, 154 pages. 1994.

Vol. 1594: M. Green, J. Murre, C. Voisin, Algebraic Cycles and Hodge Theory. Torino, 1993. Editors: A. Albano, F. Bardelli. VII, 275 pages. 1994.

Vol. 1595: R.D.M. Accola, Topics in the Theory of Riemann Surfaces. IX, 105 pages. 1994.

Vol. 1596: L. Heindorf, L. B. Shapiro, Nearly Projective Boolean Algebras. X, 202 pages. 1994.

Vol. 1597: B. Herzog, Kodaira-Spencer Maps in Local Algebra. XVII, 176 pages. 1994.

Vol. 1598: J. Berndt, F. Tricerri, L. Vanhecke, Generalized Heisenberg Groups and Damek-Ricci Harmonic Spaces. VIII, 125 pages. 1995.

Vol. 1599: K. Johannson, Topology and Combinatorics of 3-Manifolds. XVIII, 446 pages. 1995.

Vol. 1600: W. Narkiewicz, Polynomial Mappings. VII, 130 pages. 1995.

Vol. 1601: A. Pott, Finite Geometry and Character Theory. VII, 181 pages. 1995.

Vol. 1602: J. Winkelmann, The Classification of Three-dimensional Homogeneous Complex Manifolds. XI, 230 pages. 1995.

Vol. 1603: V. Ene, Real Functions – Current Topics. XIII, 310 pages. 1995.

Vol. 1604: A. Huber, Mixed Motives and their Realization in Derived Categories. XV, 207 pages. 1995.

Vol. 1605: L. B. Wahlbin, Superconvergence in Galerkin Finite Element Methods. XI, 166 pages. 1995.

Vol. 1606: P.-D. Liu, M. Qian, Smooth Ergodic Theory of Random Dynamical Systems. XI, 221 pages. 1995.

Vol. 1607: G. Schwarz, Hodge Decomposition – A Method for Solving Boundary Value Problems. VII, 155 pages. 1995.

Vol. 1608: P. Biane, R. Durrett, Lectures on Probability Theory. VII, 210 pages. 1995.

Vol. 1609: L. Arnold, C. Jones, K. Mischaikow, G. Raugel, Dynamical Systems. Montecatini Terme, 1994. Editor: R. Johnson. VIII, 329 pages. 1995.

Vol. 1610: A. S. Üstünel, An Introduction to Analysis on Wiener Space. X, 95 pages. 1995.

Vol. 1611: N. Knarr, Translation Planes. VI, 112 pages. 1995.

Vol. 1612: W. Kühnel, Tight Polyhedral Submanifolds and Tight Triangulations. VII, 122 pages. 1995.